GOVERNMENTAL RESPONSES to NATURAL DISASTERS in the U.S.

GOVERNMENTAL RESPONSES to NATURAL DISASTERS in the U.S.

A DOCUMENTARY HISTORY

John R. Vile
EDITOR

TALBOT
PUBLISHING
Clark, New Jersey

ISBN 978-1-61619-580-9

TALBOT PUBLISHING
AN IMPRINT OF
THE LAWBOOK EXCHANGE, LTD.
33 Terminal Avenue
Clark, New Jersey 07066-1321

Please see our website for a selection of our other publications and fine facsimile reprints of classic works of legal history:
www.lawbookexchange.com

Library of Congress Cataloging-in-Publication Data

Names: Vile, John R., author.
Title: Governmental responses to natural disasters in the U.S. / John R. Vile.
Description: Clark, NJ : Talbot Publishing, an imprint of the Lawbook Exchange, Ltd, [2018] | Includes bibliographical references and index.
Identifiers: LCCN 2018026855 | ISBN 9781616195809 (hardcover : alk. paper)
Subjects: LCSH: Natural disasters--Law and legislation--United States. | Disaster relief--Law and legislation--United States. | Emergency management--Law and legislation--United States. | Natural disasters--Government policy--United States.
Classification: LCC KF3750 .V55 2018 | DDC 344.7305/34--dc23
LC record available at https://lccn.loc.gov/2018026855

Printed in the United States of America on acid-free paper

Table of Contents

Timeline of Events*

79	Volcanic eruptions destroy the cities of Pompeii and Herculaneum in Italy
1665	Bubonic plague (Black Death) transmitted by fleas, sweeps through London, England killing more than 100,000 people
1666	Fire wipes out most of the homes and businesses in London, England
1700	Massive Cascadia earthquake hits near modern Seattle creating massive tsunami that was "rediscovered" in the 1980s
1755	A massive earthquake hits Lisbon, Portugal, killing thousands and sparking philosophical reflection
1776	Second Continental Congress adopts Declaration of Independence declaring U.S. independence from Great Britain
1781	Congress adopts the Articles of Confederation
1783	Revolutionary War ends with victory for former American colonies
1787	Constitutional Convention meets in Philadelphia and drafts U.S. Constitution
1789	Constitution goes into effect. Congress proposes the Bill of Rights
1790	Britain's George III extends aid to Caribbean colonies hit by hurricane
1791	States ratify the Bill of Rights

* Not all events in the text are listed here.

1803 Congress suspends bond collection for victims of Portsmouth, New Hampshire fire

1811-1812
New Madrid earthquake hits near New Madrid, Missouri and is felt throughout much of the country

Richmond Theatre fire kills 72 people

1845 A great famine caused by a potato blight in Ireland causes numerous citizens to emigrant, many of whom came to the U.S.

1848 The crops of Mormon pioneers in Salt Lake Valley are saved from crickets that are devouring their crops after seagulls eat them

1849-1850
Congress adopts series of acts turning over swamplands to state governments

1861 U.S. Civil War begins

1865 Civil War ends, Abraham Lincoln is assassinated and Congress establishes Freedmen's Bureau

States ratify Thirteenth Amendment, abolishing slavery

1868 Fourteenth Amendment seeks to guarantee equal protection and due process to all citizens

1871 The Peshtigo Fire in Michigan's Upper Peninsula leads to more than 1,000 deaths

Chicago, Illinois fire leaves thousands of people homeless

1878 Congress adopts Posse Comitatus Act

1884 Congress debates aid to flood victims along the Ohio River and its tributaries

1887 Grover Cleveland vetoes Texas Seed Bill

1889 A bursting dam leads to the Johnstown flood in Pennsylvania, which leads to changes in liability law

1893 Cheniere Caminada Hurricane kills over 2000 in Louisiana

A hurricane hits Savannah, Georgia and kills about 2000 people

1898 Sinking of the battleship Maine in Havana serves as a spark to the Spanish-American War

1900 A hurricane hits Galveston, Texas in what is believed to be the single deadliest natural disaster in U.S. history, killing about 8,000 people

American National Red Cross is chartered

1903 Fire at "fireproof" Iroquois Theater in Chicago kills 600 of 1900 attending

1906 An earthquake and accompanying fires devastates the city of San Francisco

1911 Triangle Shirt Waist Factory fire kills over 100 immigrants and helps prod states to develop worker's compensation laws and other protections

1912 The "unsinkable" *Titanic* strikes an iceberg and sinks, killing almost 1,500 people

1913 There is a statewide flood in Ohio

1917 Congress adopts Flood Control Act

1925 Tornadoes in Missouri, Illinois, and Indiana result in more than 650 deaths

1927 Great Mississippi Flood displaces many people from their homes

1928 Okeechobee Hurricane hits Puerto Rico and Florida

1929 Stock market crash leads to the Great Depression

1930s Dust storms swept U.S. and Canadian Prairies

1932 Franklin D. Roosevelt is elected president and proposes New Deal programs

1933 Tennessee Valley Authority (TVA) is established to control flooding and encourage economic development in the region

1936 Tornadoes his several southern states, and are particularly deadly in Tupelo, Mississippi, and Gainesville, Georgia

1936 Massive flood hits Pittsburgh, Pennsylvania
United States v. Butler gives an expansive reading of the taxing and spending clause

1937 The Ohio River and its tributaries flood leading to widespread loss of life and property damage

The Hindenburg zeppelin crashes in Lakehurst, New Jersey

1941 Japanese attack on Pearl Harbor, Hawaii leads to U.S. entry into World War II

1945 Atomic bombs dropped on Hiroshima and Nagasaki, Japan lead to an end of World War II

1947 An explosion aboard a ship with ammonium nitrate in Texas City leads to more than 500 deaths

1950 Congress adopts Civil Defense Act

1957 Hurricane Audrey hits Louisiana and Texas

1960 Two passenger planes collide over New York City killing 134 passengers and crew and 6 people on the ground

1964 Massive 9.2 earthquake strikes close to the coast of Prince William Sound, Alaska and causes tsunami

1965 Hurricane Betsy damages Bahamas, Florida, and Louisiana

1968 Congress adopts National Flood Insurance Act

1969 Hurricane Camille hits the South

1972 Congress adopts Coastal Zone Management Act

1973 Congress adopts Robert T. Stafford Disaster Relief and Emergency Assistance Act

1974 Numerous tornadoes hit the U.S.

1979 Partial meltdown at Three Mile Island Nuclear Plant in Pennsylvania leads to widespread fears

Carter creates the Federal Emergency Management Agency (FEMA)

1980 Mount St. Helens erupted in southwestern Washington

1989 Loma Prieta earthquake hits the San Francisco Bay area

Exxon Valdez oil tanker ran aground and spills millions of gallons that foul the coast foff Prince William Sound, Alaska for 150 miles

1992 Hurricane Andres hits Louisiana and Florida
Lucas v. South Carolina Coastal Council applies takings clause to coastal regulations

1993 Bomb explodes beneath the World Trade Center in New York City

1995 Bomb by home-grown terrorist destroys the P. Murrah Federal building in Oklahoma City killing 168 people and injuring many more

2001 Terrorist attacks on the Twin Towers in New York and on the Pentagon kill over 2000 people and lead to the establishment of the Department of Homeland Security (DHS)

2005 Hurricane Katrina hits New Orleans and leads to widespread flooding and loss of life. (August 23-31)

Hurricane Rita hits the South resulting in 120 fatalities. (September 18-26)

2011 Deep Water Horizon rig oil spill off coast of Louisiana discharges more than 210 million gallons of oil into the Gulf of Mexico

More than 350 tornadoes hi throughout the East, South, and Midwest

2012 Shooting at Sandy Hook Elementary School in Connecticut results in 28 fatalities

Hurricane Sandy causes widespread damage in the Eastern U.S.

2016 Shooting in Orlando, Florida nightclub results in 49 deaths and raises renewed fears of terrorism

Hurricane Matthew results in flooding and other damage in the southeast

2017 Donald Trump, who questioned global warming, becomes 45th U.S. President. On June 1, he announced that U.S. would cease participation in the 2015 Paris Agreement on climate change

About 188,000 people are evacuated as the Oroville Dam in California, the nation's tallest, threatened to burst

Hurricane Harvey (category 4 hurricane) leads to massive flooding in Houston, Texas and surrounding areas

In September, Hurricane Maria subsequently inflicts widespread devastation on Puerto Rico
Thomas Fire in Southern California in December is largest in state's history since 1932

2018 National Centers for Environmental Information (NCEI) indicate that 2017 disasters were the costliest on record

Introduction

This book is designed to remedy what I believe to be a lacuna within existing scholarship. After speaking with a publisher, I began thinking about compiling a book that would record the history of the Second Amendment and its applications, only to find that the field was fairly well occupied. By contrast, when I began researching disaster aid, I discovered that although there were plenty of references to history and numerous accounts of individual catastrophes, I found it difficult to find a compilation of original sources that would adequately place modern issues in context. I had just completed such a compilation of documents on issues related to immigration and citizenship, and this seemed to present another such opportunity.

Rarely does a month go by without notice of a major natural catastrophe in the United States and in the rest of the world. Although many accounts of disaster are related to war, for example, that in Syria or Yemen, I have followed other scholarship in this area by largely confining readings in this book to issues related to natural disasters, which are, however, sometimes precipitated or aggravated by human conduct. When I took my first full-time professorial job in Louisiana, my wife and I faced a hurricane in our first week, and we lived through many more during our 12-year stay there. Since living in Tennessee, a number of neighboring communities have been struck by tornadoes. As I was writing this book, Hurricane Matthew caused widespread damage in the Southeast, and wild fires swept through Gatlinburg, Tennessee causing widespread destruction of property and some loss of life.

The more I read about disasters, the more I realized that while disasters are hardly new, most existing writings focus on the latest catastrophe or set of catastrophes. I believe that we might well get a wider understanding of the subject by attempting to understand contemporary institutions and events within an historical context. I have attempted to do this by reaching back into Biblical and classical history for early readings and continuing throughout U.S. history to see how current laws, institutions, and procedures have evolved.

Almost any periodization is somewhat arbitrary, but I have divided this book into seven time periods, each of which shared similar concerns. Where possible I have included readings from each period from all of the three branches of the national government. In addition to the philosophical reflections that largely dominate the readings in the first part of this book, most successive parts include combinations of presidential addresses and orders, state and congressional actions, and judicial decisions as well as occasional charters and other relevant documents, which now exceed 75 in number.

PART I Historical and Philosophical Context

Part I of this book somewhat ambitiously begins with stories from the Bible, continues with reflections from the Renaissance and Age of Enlightenment, and concludes with a reading from economist and moralist Adam Smith. Stories from the Biblical book of Genesis show that the ancients often interpreted natural disasters as divine punishment for sins. In two stories that I have included, however, those of Noah and Abraham, God chooses to spare the righteous. In the case of Noah, he did so by instructing him to build an ark, which would also preserve wildlife. In the case of Abraham, he warns his nephew Lot and his family of the impending destruction of Sodom and Gomorrah so they could flee to a neighboring town.

The most fascinating story, however, is probably that of Joseph. After his borthers sell him into slavery in Egypt, Joseph interprets the Pharaoh's dream to forecast seven years of bountiful harvests followed by seven years of famine. The Pharaoh puts Joseph in charge of a program to store grain in the good years to prepare for the bad. Moreover, in so doing, he is acting as an agent of the government. His actions end up not only saving the Egyptians, while arguably subjecting them to greater control by the Pharaoh, but also his own family in Canaan. The story is thus an early, and effective, example of disaster preparedness and mitigation, albeit one that may have increased the power of ruling authorities.

Moving to the New Testament, I have included Jesus's parable of the Good Samaritan, which stresses that neighborliness involves crossing cultural boundaries to help fellow individuals in need. This message is continued in St. Paul's epistles where he urges members of the church to consider themselves to be members of a common body, sharing not only one another's joys but also their sufferings, which he balances against the admonition to the lazy that "if a man does not work, he should not eat." Not long after, Roman historians would record how emperors and members of the patrician class often provided relief to those who were suffering from natural disasters like the volcano that destroyed Pompeii and Herculaneum in 79 AD. Philanthropists and modern entertainers still sometimes expend special efforts in relieving suffering caused by natural catastrophes.

As we move to the writings of Niccolo Machiavelli, we see a classic conflict between fatalists, who essentially believe that what will be will be, versus a man of both reflection and action who seeks to bend fortune through prior preparation. Tellingly, Machiavelli believes that men's ability to control physical floods through dams and dykes is an indication of the manner in which they can also use institutions to control political developments.

Later British social contract philosophers sought to understand government by portraying what humankind would be like in a state of nature without it. Thomas Hobbes painted a grim picture of such a stateless state in which the life of man would be "solitary, poor, nasty, brutish, and short." Although John Locke believed that men would still be obligated to respect natural law, he acknowledged that they would still want to escape the state of nature because such law would not be settled or known, and it would lack adequate interpreters and enforcers. Just as humans would leave such a state of nature and join a social contract to protect their lives, liberties, and property, so too, modern men and women seek to avoid and repair the chaos that often emerges after natural disasters strike. Emerich de Vattel describes how international law sometimes extends obligations to extend aid beyond one's own homeland, while Voltaire ridicules those who believe that interfering with natural disasters, like the great earthquake in Lisbon, is somehow against God's will.

I end this section with a selection from Adam Smith, who is best known for his *Wealth of Nations,* which was published the year that the American Revolution began. Taken from his *Theory of Moral Sentiments,* this selection shows how Smith understood how difficult it can be to sympathize with individuals who are out of sight. Modern television broadcasts make it increasingly difficult to be unaware, or unconcerned about, disasters throughout the world, but especially in our own nation. News that once took weeks or months to arrive is now often broadcast to our living rooms as it is happening. Such news puts increased pressures on public officials to act and to act quickly and effectively. Such pressures also sometimes leads to donor fatigue, especially in cases of war where it is often very difficult for private individuals to know how to intervene effectively on behalf of the innocent.

PART II Constitutional Foundations and Early U.S. History

Part II of this book begins shifting the focus to America and specifically to America's Founding period, which is one of my favorite subjects (give a man a hammer, and he will see everything as a nail!). As a professor of constitutional law, I am particularly interested in the manner in which the Constitution and other documents have shaped our understandings of the role of government. Because the Declaration of Independence appears to make no mention of natural disasters (although it does accuse the king of war atrocities), I begin with America's first constitution, the Articles of Confederation, and cite phrases that would be echoed in the U.S. Constitution.

I go on to cite the Preamble to the Constitution with its concern for the "general welfare," the clauses that provide the most direct support for governmental disaster aid (most notably the taxing and spending and commerce clauses), and some notice of the institutions that are most likely to carry it out. Although the Constitution significantly strengthened the national government, it retained a federal system that left major responsibilities to the state. Much of the literature of disaster aid attempts to assign appropriate responsibilities to each of these levels of government and to stress the need for cooperation among them.

I continue by showing how King George III decided to provide governmental aid to the Caribbean after a hurricane, much as the British had responded to an earthquake in South Carolina in 1740. As befits his Puritan heritage, John Adams's Proclamation of a Day of Fasting, Humiliation, and Prayer further demonstrates how early Americans continued to reflect earlier ideas associating disasters with God's wrath. I also include a reading from the British economist Thomas Malthus, to serve as a reminder of the continuing influence of European political thought on the Americas and dire predictions that might indicate that all relief efforts simply postpone the inevitable. I leave it to readers to draw parallels between Malthus's thoughts and modern harbingers of "the population bomb," the inevitable effects of global warming, or portents of other world-wide catastrophes like being hit by a large meteorite.

The year 1803 apparently marks the date of the first form of federal aid to a state in the wake of a disaster, in this case a fire in Portsmouth, New Hampshire. The brevity of this law makes a studied (and welcome) contrast to the complexity of much of the legislation that Congress has passed in the twentieth and twenty-first centuries. The Insurrection Act demonstrates the manner in which the government stood prepared to quell insurrections, while St. George Tucker makes an argument to justify the use of federal power, despite a veto of President James Monroe's veto of the Cumberland Road Bill, which is also included. Justice Joseph Story provides one of the best argued interpretations of the taxing and spending clause, while a selection from the Swamp Lands Act of 1849-1850 reminds us that actions taken in one century to alleviate problems are often considered in a later century as contributors to them.

PART III The Civil War and Reconstruction

The period from 1865 to 1877 was necessarily dominated less by concerns over natural catastrophes than by the Civil War and the subsequent congressional reconstruction of the South. This war brought more dislocation than any war in U.S. history, leaving much of the population, including newly emancipated freedmen, extremely vulnerable to economic downturns and natural catastrophes.

Immediately after the war, the United States adopted three amendments designed to assure that slavery did not return and to guarantee the rights of all citizens. Although the country long lagged in guaranteeing the same due process and equal protection rights to African-Americans that it guaranteed to others, these amendments, which included congressional enforcement clauses, would set the stage for wider federal engagement in all areas of national life.

One indication of this new role was the creation of the Freedmen's Bureau, whose renewal President Andrew Johnson sought to veto. The nation continued to question whether it was appropriate for governments to give assistance that private charitable actions had traditionally provided. The section ends with a variety of court decisions on the subject as well as a limit on the use of federal troops (the Posse Comitatus Act of 1878), which was largely adopted in reaction to congressional Reconstruction and which continues into effect today.

PART IV From the End of the Nineteenth Century Through World War I

Individuals in the late nineteenth century were both shocked and fascinated by Charles Darwin's articulation of the theory of evolution, which many began applying to governmental actions. The selection from William Graham Sumner is designed to reflect the philosophy of Social Darwinism, which eschewed governmental intervention but, at least in Sumner's case, was willing to indulge private charity. Grover Cleveland would reflect this philosophy when he stated, in vetoing, a bill designed to provide seed to drought-stricken farmers, that "Though the people support the Government the Government should not support the people." As congressional debates over aid to Ohio flood victims demonstrate, however, others either favored interpreting the Constitution capaciously enough to provide such aid or suggested that necessity trumped constitutional niceties in such circumstances.

The chartering of the American National Red Cross demonstrated the continuing reliance on private aid during this period. In the meantime, governments changed laws, often in the wake of catastrophes, like the Johnstown Flood of 1889 and the Shirt Waist Factory Fire of 1911, to provide greater legal recourse to those who were hurt in part through the actions of others. The Johnstown Flood was a particularly good example of the manner in which private actions, and inactions (negligence), could sometimes trigger natural catastrophes.

The San Francisco Earthquake of 1905 provided yet another example of the variety of catastrophes that could affect a nation that now spanned a continent, while the mayor's shoot-to-kill order reveals the underlying fear of chaos that often accompanies disasters. Throughout this period, the government increasingly found itself engaged in prevention efforts.

PART V The New Deal and World War II

It is often difficult to categorize catastrophes. Although this book focuses largely on natural disasters, the effects of economic downturns can often be as widespread and as significant as hurricanes or tornadoes. The stock market crash of 1929 helped plunge America and the world into a Great Depression, the continuing economic impact of which continued for the next ten years. This Depression has affected understandings of the role of government to the present day.

In 1932, the nation elected Franklin D. Roosevelt who swept into office with a wide array of national governmental programs that he labeled the New Deal and that insinuated government into almost every area of American life. The Soil Conservation Act sought to combat the ravages of the dust bowl and the widespread migration that it spurred, while the Tennessee Valley Authority sought both to control flooding and to generate cheap electricity to revitalize an economically disadvantaged area. Although the U.S. Supreme Court initially overturned a number of these initiatives, in time it used an expansive constitutional interpretation to endorse most of them.

Roosevelt created an Office of Civilian Defense, but faced continuing divisions, among those, like director Fiorello La Guardia, who focused on foreign attacks and those, like assistant director Eleanor Roosevelt, who was more concerned with "social defense." As World War II approached, the nation turned increasingly to national preparedness for national security against foreign foes. The Japanese air attack on Pearl Harbor on December 7, 1941 – what Roosevelt called "a date that will live in infamy" -- led Roosevelt to turn his attention from the national economy to the national defense. Still, through the end of his administration, Roosevelt continued to stress the need for the government to provide security against economic as well as natural catastrophes. Although he did not succeed in incorporating social and economic rights as written constitutional guarantees, he emphasized "freedom from fear" and introduced a broad vision of government that both President Lyndon Johnson's ambitious plans for a Great Society as well as in the recent plans for national health care have subsequently reflected.

PART VI The Cold War and Increasing Institutionalization of Disaster Responses

No sooner had World War II ended than America found itself in a struggle against international communism as embodied in the Soviet Union and "Red" China, which both came to possess nuclear weapons. The next few decades would find the nation often committing more resources to preparing for a nuclear war, which thankfully never came, than in preparing for more traditional natural disasters. A number of Federal Civil Defense Acts reflected this commitment.

As time passed, Congress did adopt an increasing number of laws designed to deal with losses from floods and other disasters. Most notably, in 1973, the Robert T. Stafford Disaster Relief and Emergency Assistance Act provided for presidential declarations of emergency that would bring about more immediate aid. Moreover, President Jimmy Carter created the Federal Emergency Management Agency (FEMA) in 1979. No matter what new programs the national government adopted, they needed to be coordinated with state and local agencies, and many laws adopted during this period attempted to flesh out this relationship.

By 1987, the United Nations had proclaimed a decade for natural disaster reduction. At home, Supreme Court opinions on the takings clause, most notably reflected in *Lucas v. South Carolina Coastal Council* (1992), sometimes seemed at odds with rules designed to regulate building in flood and other disaster-prone areas that the government sometimes itself encouraged by subsidizing insurance for such regions.

PART VII 9/11, Hurricane Katrina, Global Warming, and the 2017 Hurricanes

Just at the previous decades were dominated by fears of nuclear war, so too, the decades after the terrorist attacks of 2001 have been dominated by fears of and preparations for terrorist attacks. This was reflected not only in President Bush's declaration of war against terrorism but also by his decision to fold the Federal Emergency Management Agency, which had been created to deal chiefly with natural disasters, into the new Department of Homeland Security, where emphasis on national security against foreign threats often overshadows concern over domestic natural disasters.

Numerous scholars have pointed out that individuals often assess risks incorrectly. Most of us are far more likely to die in car accidents than in terrorist attacks, and yet the nation devotes far more resources to defending against the latter than the former. The images from 9/11 are likely to continue to influence threat assessments in the near future and may direct resources away from more likely natural disasters.

One of the greatest natural disasters of this time period was Hurricane Katrina, which struck the Gulf Coast in 2005 and resulted in widespread flooding and loss of life, especially in New Orleans. Television reporters were on hand to report mayhem, often in highly exaggerated accounts, and what appeared to be widespread governmental incompetence in dealing with the crisis. The event continues to be a focus of much subsequent commentary.

One lesson learned, which has since been embodied in legislation, is that individuals often would not evacuate during emergencies without their pets. Subsequent disasters, most notably Hurricane Sandy, demonstrated that Congress had not appropriated enough money to take care of yearly disaster expenses and often required budget-busting individualized appropriations to deal with unanticipated catastrophes.

As critics of governmental efforts have increased, some individuals have begun to refocus on efforts of religious organizations and other private relief agencies like the Salvation Army whose members could often provide the emotional and spiritual support that secular governments could not. The selection from the bylaws of the National Voluntary Organizations Active in Disasters (NVOAD), like the Emergency Management Assistance Compact (1994) reported in the previous section, is designed to show how such organizations seek to coordinate their own relief efforts.

As the nation entered a new century, there was increasing concern that the rising severity of natural disasters, both at home and abroad, might be related to the effects of mankind's increasing use of fossil fuels, which released carbon dioxide in the atmosphere and led to increases in temperatures. As the nation ushered in a new administration, politicians continue to debate the severity of the crisis and what, if any, governmental actions are needed to deal with it.

Audiences

I have designed this book for public, university, and law libraries as well as for instructors in classes dealing with emergency management, disaster preparedness, and related fields. I think it would be especially appropriate as a book of primary readings to supplement existing textbooks. This is not a handbook for those writing laws or responding to disasters, but I think that such individuals might well benefit from looking beyond both the most recent catastrophe or the next anticipated one. I would also be pleased if general readers find this book an appropriate tool for reflecting on a major issue of public policy.

Themes

Because this is a book of readings rather than an extended text, I have largely left readers with the task of connecting individual essays and developing coherent themes, but this book certainly raises important questions. They include:

• The degree to which a Constitution designed largely to protect against governmental actions that restrict freedoms can be utilized to protect citizens against natural disasters, which often undermine their social and economic well-being.

• The degree to which aid programs may funnel more moneys to those with higher incomes—home owners who lose their beach houses for example—than to those with fewer resources, like renters.

• The degree to which immigration and naturalization policies should be designed to take care of individuals who are displaced by war or natural disasters and the degree to which such problems might better be handled through foreign aid and other policies.

• The degree to which it is wise to devise institutions and link issues of safety against natural disasters with concerns over national security.

• The degree to which preparations for, and responses to, routine emergencies should be tied to those for greater disasters.

• The proper balance of state and federal actions and private and governmental actions and the best way to provide the former without deterring or undermining the latter.

• The degree to which government rebuilding programs can lead to moral hazard by discouraging individuals from engaging in behavior and taking the kinds of precautions they might were they to bear the entire costs.

• The appropriate response to reports of global warming and the degree to which the United States can effectively participate in international treaties without undermining national sovereignty.

I would not be able to give complete answers to many of these questions, and it is unlikely than any two readers will come to the same conclusions. I have designed this book, however, to be a good starting point for such reflections, particularly when used in conjunction with textbooks that focus specifically on disaster preparedness and relief.

Resources

I am a firm believer in the value of knowing and understanding primary texts, and this book reflects this belief. I have attempted to reduce the intimidation factor in reading original sources both by including such diverse sources in a single place and by editing them to manageable lengths, which should be accessible to most high school and college students who might have difficulty in locating them and identifying key sections on their own.

In addition to this introduction, I have included brief previews that identify individual authors and explain the background and the significance of each individual essay. As in the case of my book on citizenship and naturalization, I have further sought to provide essays from all three branches of the national government, as well as occasional reminders of the role of religious groups and other private charitable organizations.

In addition to this introduction and a detailed table of contents, the book includes a timeline, a short bibliography, and an index. Readers will see that, apart from the opening section, I concentrate almost solely on the United States although I think the lessons from the American experience are often applicable elsewhere.

Acknowledgements

I am pleased to acknowledge those who have helped me with this project. They include the following: Susan Lyons, who helped file increasing mounds of documents and essays and proofread; Miranda Dotson, a student aide who valiantly tracked down and copied relevant parts of the *Congressional Record*; Matthew Hasan, another student aide who proofed this manuscript; reference librarians and members of MTSU's Interlibrary Loan Office who responded to numerous requests for help; and colleagues and family members who continue to demonstrate patience in listening to me explicate my latest ideas while sometimes offering helpful suggestions of their own. I am particularly indebted to other scholars in this field, whose work I have relied upon heavily and to supportive administrators at my home institution. I am especially grateful to have a job as Dean of the University Honors College at Middle Tennessee State University, whose extraordinary students and staff are incredible inspiring partners.

John R. Vile
Middle Tennessee State University

Dedicated to the nation's first responders including private citizens and members of religious and charitable organizations who embody the Biblical spirit of good neighborliness.

GOVERNMENTAL RESPONSES to NATURAL DISASTERS in the U.S.

PART I

Historical and Philosophical Context

And the waters prevailed upon the earth
an hundred and fifty days

THE STORY OF NOAH

Genesis 7:1-24

Introduction

The Biblical book of Genesis records that God created the earth and that it was good. The story of Adam and Eve, the first humans, also teaches that human sinfulness resulted in a fallen world, symbolized by thorns and thistles, by the necessity for human toil, and by death.

Of all the natural disasters recorded in human history, none is as epic as the story of the flood that covered the earth in the days of Noah. Warned by God that He was going to punish the earth for its sins, God instructed Noah to construct an ark to protect himself and his family. Perhaps as a way of showing human dependency on, and/or stewardship of, the surrounding natural world, Noah was also given responsibility for sheltering wildlife so that they could repopulate the earth once the flood ended. At the end of the ordeal, God promised never to destroy the earth again by water, a promise symbolized by the rainbow.

Like other accounts from ancient times, this story both ties certain natural disasters to God's wrath and suggests that while some such disasters cannot be prevented, they can nonetheless be mitigated through human preparation. Floods, often the results of hurricanes, cyclones, and tsunamis, continue to be among the most devastating of modern disasters.

1 And the LORD said unto Noah, Come thou and all thy house into the ark; for thee have I seen righteous before me in this generation.

2 Of every clean beast thou shalt take to thee by sevens, the male and his female: and of beasts that are not clean by two, the male and his female.
3 Of fowls also of the air by sevens, the male and the female; to keep seed alive upon the face of all the earth.
4 For yet seven days, and I will cause it to rain upon the earth forty days and forty nights; and every living substance that I have made will I destroy from off the face of the earth.
5 And Noah did according unto all that the LORD commanded him.
6 And Noah was six hundred years old when the flood of waters was upon the earth.
7 And Noah went in, and his sons, and his wife, and his sons' wives with him, into the ark, because of the waters of the flood.
8 Of clean beasts, and of beasts that are not clean, and of fowls, and of every thing that creepeth upon the earth,
9 There went in two and two unto Noah into the ark, the male and the female, as God had commanded Noah.
10 And it came to pass after seven days, that the waters of the flood were upon the earth.
11 In the six hundredth year of Noah's life, in the second month, the seventeenth day of the month, the same day were all the fountains of the great deep broken up, and the windows of heaven were opened.
12 And the rain was upon the earth forty days and forty nights.
13 In the selfsame day entered Noah, and Shem, and Ham, and Japeth, the sons of Noah, and Noah's wife, and the three wives of his sons with them, into the ark;
14 They, and every beast after his kind, and all the cattle after their kind, and every creeping thing that creepeth upon the earth after his kind, and every fowl after his kind, every bird of every sort.
15 And they went in unto Noah into the ark, two and two of all flesh, wherein is the breath of life.
16 And they that went in, went in male and female of all flesh, as God had commanded him: and the LORD shut him in.
17 And the flood was forty days upon the earth; and the waters increased, and bare up the ark, and it was lift up above the earth.

18 And the waters prevailed, and were increased greatly upon the earth; and the ark went upon the face of the waters.

19 And the waters prevailed exceedingly upon the earth; and all the high hills, that were under the whole heaven, were covered.

20 Fifteen cubits [a cubit is generally believed to have been about 18 inches] upward did the waters prevail; and the mountains were covered.

21 And all flesh died that moved upon the earth, both of fowl, and of cattle, and of beast, and of every creeping thing that creepeth upon the earth, and every man:

22 All in whose nostrils was the breath of life, of all that was in the dry land, died.

23 And every living substance was destroyed which was upon the face of the ground, both man, and cattle, and the creeping things, and the fowl of the heaven; and they were destroyed from the earth: and Noah only remained alive, and they that were with him in the ark.

24 And the waters prevailed upon the earth an hundred and fifty days.

God remembered Abraham, and sent Lot
out of the midst of the overthrow

THE DESTRUCTION OF SODOM AND GOMORRAH

Genesis 19:24-29

Introduction

The Old Testament, which Christians share in common with Jews and Muslims, details many accounts of God's judgment on individuals and nations who transgressed His will. In this passage from the book of Genesis, God decides to destroy the wicked cities of Sodom and Gomorrah but, balancing judgment with mercy, seeks to spare the family of Lot, a nephew of Abraham, who was a just man. If fire and brimstone are designed to describe a natural phenomenon, it might be a volcano.

24 Then the Lord rained upon Sodom and upon Gomorrah brimstone and fire from the Lord out of heaven;

25 And he overthrew those cities, and all the plain, and all the inhabitants of the cities, and that which grew upon the ground.

26 But his wife looked back from behind him, and she became a pillar of salt.

27 And Abraham gat up early in the morning to the place where he stood before the Lord:

28 And he looked toward Sodom and Gomorrah, and toward all the land of the plain, and beheld, and, lo, the smoke of the country went up as the smoke of a furnace.

29 And it came to pass, when God destroyed the cities of the plain, that God remembered Abraham, and sent Lot out of the midst of the overthrow, when he overthrew the cities in the which Lot dwelt.

King James Version (KJV)

And the famine was over all the face of the earth: and Joseph opened all the storehouses, and sold unto the Egyptians

---

THE STORY OF JOSEPH AND FAMINE

Genesis 41:1-57

Introduction

One of the most profound stories in the Old Testament (Hebrew Scripture) is the story of Joseph, which the opening book of Genesis recorded. The brothers of Joseph, the eleventh of twelve sons and the favorite of his father Jacob, sold into slavery in Egypt and told their father that he had been killed by wild beasts. Joseph was an Horatio Alger of the patriarchs who rose in rank and was known for his ability to interpret dreams but faced setbacks including unjust imprisonment, where he is at the beginning of this passage.

Pharaoh has a dream that Joseph correctly interprets as predicting seven good years of agricultural plenty followed by seven years of famine. Joseph urges Pharaoh to prepare for this pending disaster by using the good years to store grain to prepare for the bad. Pharaoh does so by appointing Joseph to lead the efforts. At least one political scientist has observed that Joseph's advice leads to centralization and the strengthening of the pharaoh's power (See Wildavsky, 1984), which was inimical to individual liberty but which also prevented mass starvation and eventually led Joseph's brothers to come from Israel in search of food. This ultimately led to family reconciliation and salvation.

Although the wording is a bit archaic, the King James Version used here would have been among the most familiar to America's Founders who were very biblically literate. Unlike some other Biblical stories, for example the story of Noah's preparations for the great flood, where natural catastrophes or military losses are attributed to God's judgment, God's involvement in this story appears largely in providing an interpretation to Joseph as a way of allowing Pharaoh to adopt policies to save his people. This salvation will extend to their Hebrew neighbors, who will migrate to Egypt and remain there until God uses Moses to deliver them from slavery under a pharaoh who no longer remembered Joseph's role in the famine alleviation effort.

Ironically, while the Jews come to Egypt to escape a natural disaster (famine), they leave after Moses calls down a number of natural disasters on the Pharaoh and his people, which demonstrate the superiority of the God of the Jews over the gods of Egypt.

1 And it came to pass at the end of two full years, that Pharaoh dreamed: and, behold, he stood by the river.
2 And, behold, there came up out of the river seven well favored kine [cows] and fat-fleshed; and they fed in a meadow.
3 And, behold, seven other kine came up after them out of the river, ill favored and lean-fleshed; and stood by the other kine upon the brink of the river.
4 And the ill favored and lean-fleshed kine did eat up the seven well favored and fat kine. So Pharaoh awoke.
5 And he slept and dreamed the second time: and, behold, seven ears of corn came up upon one stalk, rank and good.
6 And, behold, seven thin ears and blasted with the east wind sprung up after them.
7 And the seven thin ears devoured the seven rank and full ears. And Pharaoh awoke, and, behold, it was a dream.
8 And it came to pass in the morning that his spirit was troubled; and he sent and called for all the magicians of Egypt, and all the wise men thereof: and Pharaoh told them his dream; but there was none that could interpret them unto Pharaoh.
9 Then spake the chief butler unto Pharaoh, saying, I do remember my faults this day:
10 Pharaoh was wroth with his servants, and put me in ward in the captain of the guard's house, both me and the chief baker:

11 And we dreamed a dream in one night, I and he; we dreamed each man according to the interpretation of his dream.

12 And there was there with us a young man, an Hebrew, servant to the captain of the guard; and we told him, and he interpreted to us our dreams; to each man according to his dream he did interpret.

13 And it came to pass, as he interpreted to us, so it was; me he restored unto mine office, and him he hanged.

14 Then Pharaoh sent and called Joseph, and they brought him hastily out of the dungeon: and he shaved himself, and changed his raiment, and came in unto Pharaoh.

15 And Pharaoh said unto Joseph, I have dreamed a dream, and there is none that can interpret it: and I have heard say of thee, that thou canst understand a dream to interpret it.

16 And Joseph answered Pharaoh, saying, It is not in me: God shall give Pharaoh an answer of peace.

17 And Pharaoh said unto Joseph, In my dream, behold, I stood upon the bank of the river:

18 And, behold, there came up out of the river seven kine, fa-tfleshed and well favored; and they fed in a meadow:

19 And, behold, seven other kine came up after them, poor and very ill favored and lean-fleshed, such as I never saw in all the land of Egypt for badness:

20 And the lean and the ill favored kine did eat up the first seven fat kine:

21 And when they had eaten them up, it could not be known that they had eaten them; but they were still ill favored, as at the beginning. So I awoke.

22 And I saw in my dream, and, behold, seven ears came up in one stalk, full and good:

23 And, behold, seven ears, withered, thin, and blasted with the east wind, sprung up after them:

24 And the thin ears devoured the seven good ears: and I told this unto the magicians; but there was none that could declare it to me.

25 And Joseph said unto Pharaoh, The dream of Pharaoh is one: God hath showed Pharaoh what he is about to do.

26 The seven good kine are seven years; and the seven good ears are seven years: the dream is one.
27 And the seven thin and ill favored kine that came up after them are seven years; and the seven empty ears blasted with the east wind shall be seven years of famine.
28 This is the thing which I have spoken unto Pharaoh: What God is about to do he showeth unto Pharaoh.
29 Behold, there come seven years of great plenty throughout all the land of Egypt:
30 And there shall arise after them seven years of famine; and all the plenty shall be forgotten in the land of Egypt; and the famine shall consume the land;
31 And the plenty shall not be known in the land by reason of that famine following; for it shall be very grievous.
32 And for that the dream was doubled unto Pharaoh twice; it is because the thing is established by God, and God will shortly bring it to pass.
33 Now therefore let Pharaoh look out a man discreet and wise, and set him over the land of Egypt.
34 Let Pharaoh do this, and let him appoint officers over the land, and take up the fifth part of the land of Egypt in the seven plenteous years.
35 And let them gather all the food of those good years that come, and lay up corn under the hand of Pharaoh, and let them keep food in the cities.
36 And that food shall be for store to the land against the seven years of famine, which shall be in the land of Egypt; that the land perish not through the famine.
37 And the thing was good in the eyes of Pharaoh, and in the eyes of all his servants.
38 And Pharaoh said unto his servants, Can we find such a one as this is, a man in whom the Spirit of God is?
39 And Pharaoh said unto Joseph, Forasmuch as God hath showed thee all this, there is none so discreet and wise as thou art:
40 Thou shalt be over my house, and according unto thy word shall all my people be ruled: only in the throne will I be greater than thou.

41 And Pharaoh said unto Joseph, See, I have set thee over all the land of Egypt.

42 And Pharaoh took off his ring from his hand, and put it upon Joseph's hand, and arrayed him in vestures of fine linen, and put a gold chain about his neck;

43 And he made him to ride in the second chariot which he had; and they cried before him, Bow the knee: and he made him ruler over all the land of Egypt.

44 And Pharaoh said unto Joseph, I am Pharaoh, and without thee shall no man lift up his hand or foot in all the land of Egypt.

45 And Pharaoh called Joseph's name Zaphnathpaaneah; and he gave him to wife Asenath the daughter of Potipherah priest of On. And Joseph went out over all the land of Egypt.

46 And Joseph was thirty years old when he stood before Pharaoh king of Egypt. And Joseph went out from the presence of Pharaoh, and went throughout all the land of Egypt.

47 And in the seven plenteous years the earth brought forth by handfuls.

48 And he gathered up all the food of the seven years, which were in the land of Egypt, and laid up the food in the cities: the food of the field, which was round about every city, laid he up in the same.

49 And Joseph gathered corn as the sand of the sea, very much, until he left numbering; for it was without number.

50 And unto Joseph were born two sons before the years of famine came, which Asenath the daughter of Potipherah priest of On bare unto him.

51 And Joseph called the name of the firstborn Manasseh: For God, said he, hath made me forget all my toil, and all my father's house.

52 And the name of the second called he Ephraim: For God hath caused me to be fruitful in the land of my affliction.

53 And the seven years of plenteousness, that was in the land of Egypt, were ended.

54 And the seven years of dearth began to come, according as Joseph had said: and the dearth was in all lands; but in all the land of Egypt there was bread.

55 And when all the land of Egypt was famished, the people cried to
Pharaoh for bread: and Pharaoh said unto all the Egyptians, Go
unto Joseph; what he saith to you, do.
56 And the famine was over all the face of the earth: and Joseph
opened all the storehouses, and sold unto the Egyptians; and the
famine waxed sore in the land of Egypt.
57 And all countries came into Egypt to Joseph for to buy corn;
because that the famine was so sore in all lands.

Go, and do thou likewise

THE PARABLE OF THE GOOD SAMARITAN

Luke 10: 25-37

Introduction

One of Jesus' best-known parables is that of the Good Samaritan, which the gospel of Luke (the third of three accounts of the life of Jesus, written by a physician who later accompanied St. Paul on his missionary journeys) records. Used to illustrate what Jesus identifies as the greatest commandment, Jesus designed the parable to help his hearers expand their understanding of who their neighbors are.

The Jewish audience to whom Jesus addressed this parable, looked down on Samaritans, who had intermarried with foreign conquerors. In this story, however, the Samaritan shows greater compassion than Jewish leaders headed to the temple.

This parable focused on individual action to a single individual in need, but the principles would obviously embody aid to groups of individuals in similar distress. Jesus did not in this parable directly address the degree to which governments should assume the role that the Good Samaritan assumes in this story.

25 And, behold, a certain lawyer stood up, and tempted him, saying, Master, what shall I do to inherit eternal life?
26 He said unto him, What is written in the law? how readest thou?
27 And he answering said, Thou shalt love the Lord thy God with all thy heart, and with all thy soul, and with all thy strength, and with all thy mind; and thy neighbour as thyself.
28 And he said unto him, Thou hast answered right: this do, and thou shalt live.

29 But he, willing to justify himself, said unto Jesus, And who is my neighbour?

30 And Jesus answering said, A certain man went down from Jerusalem to Jericho, and fell among thieves, which stripped him of his raiment, and wounded him, and departed, leaving him half dead.

31 And by chance there came down a certain priest that way: and when he saw him, he passed by on the other side.

32 And likewise a Levite, when he was at the place, came and looked on him, and passed by on the other side.

33 But a certain Samaritan, as he journeyed, came where he was: and when he saw him, he had compassion on him,

34 And went to him, and bound up his wounds, pouring in oil and wine, and set him on his own beast, and brought him to an inn, and took care of him.

35 And on the morrow when he departed, he took out two pence, and gave them to the host, and said unto him, Take care of him; and whatsoever thou spendest more, when I come again, I will repay thee.

36 Which now of these three, thinkest thou, was neighbour unto him that fell among the thieves?

37 And he said, He that shewed mercy on him. Then said Jesus unto him, Go, and do thou likewise.

Even a father's surpassing love

C. Suetonius Tranquillus
THE LIVES OF THE TWELVE CAESARS

Introduction

Many modern visitors to Italy learn of an ancient catastrophe when they visit Pompeii and Herculaneum, which were covered by volcanic ash in an eruption in 79 AD. Pliny witnessed and described this disaster. Titus Flavius Caesar Vespasianus Augustus (39-81 AD) was the reigning emperor during this time.

The passage below indicates that Titus gained popular favor by attempting to provide aid to the victims of this and other catastrophes. Given the emperor's role at the time, it is difficult to distinguish his actions as a citizen from those as a ruler. As someone whom the people would deify after his death, it is not surprising to see that part of his role involved that of making sacrifices. Later Christian emperors sometimes further accompanied provision of post catastrophic aid with displays of "feelings of humility and mourning." Thus Theodosius walked barefoot in the Senate after an earthquake in 447, and Justinian dressed in rags and wept after the earthquake of 526 in Antioch and put aside his crown after an earthquake in Constantinople in 557.[1]

There were some dreadful disasters during his reign, such as the eruption of Mount Vesuvius in Campania, a fire at Rome which continued three days and as many nights, and a plague the like of which had hardly ever been known before. In these many great calamities he showed not merely the concern of an emperor, but a

---

[1] Christopher M. Higgins, *Popular and Imperial Responses to Earthquakes in the Roman Empire*, MA. Thesis, Ohio University, June 2009, p. 111.

father's surpassing love, offering consolation in edicts, and lending aid so far as his means allowed. He chose commissioners by lot from among the ex-consults for the relief of Campania; and the property of those who lost their lives by Vesuvius and had no heirs left alive he applied to the rebuilding of the buried cities. During the fire in Rome he made no remark except "I am ruined," and he set aside all the ornaments of his villas for the public buildings and temples, and put several men of the equestrian order in charge of the work, that everything might be done with the greater dispatch. For curing the plague and diminishing the force of the epidemic there was no aid, human or divine, which he did not employ, searching for every kind of sacrifice and all kinds of medicines.

Source: TC. Suetonius Tranquillus, *The Lives of the Twelve Caesars,* published in the Loeb Classical Library, 1914. Found at http://penelope.uchicago.edu/Thayer/E/Roman/Texts/Suetonius/12Caesars/Titus*.html. Note indicates that the text is in the public domain.

And whether one member suffer, all the members suffer with it

---

St. Paul
EPISTLES

Introduction

One of the images that St. Paul and other Christian writers used to describe the church was that of a body. In I Corinthians, which Paul wrote to the early church in Corinth, Greece, Paul said that, like a human body, the church, which is often called "the body of Christ," consisted of many members with complementary gifts and functions, all of which were designed to minister to the body as a whole. Just as an entire body would suffer when one of its members suffered, so too members of the church should support one another in their distress.

A large body of political literature subsequently developed throughout the middle ages and the early modern period that likened government to a body, "the body politic." Most of this literature appeared directed to indicating that a government, like a body, needed to be well-ordered, often stressing the need for a head, often a monarch. [See Andrei Constantin Salavastru, "The English 'Body Politic': The Role of a Metaphor in Renaissance English Political Theory, in *Communication, Context, Interdisciplinarity Studies and Articles, Volume III Section: History,* Julian Boldea coordinator. "Petru Maior" University Press, Tirgu Mures, pp. 336-345.]

To the extent that such analogies might also harken back to the image of the church, however, it would stand to reason that the body as a whole should look out after individual members when they were in distress. This conception might well be in tension with interpretations that suggest that the unity of the state was not an organic unity like that of a body or even a church, but that the common good resulted from the work of the "invisible hand" of the marketplace, as in laissez-faire economics, or simply from individuals pursuing their individual interests.

The doctrines of separation of powers and checks and balances as embodied in the U.S. Constitution suggest that individuals pursuing the interest of their office might serve to check abuses of power by members of the other branches. Such doctrines might further suggest that private charity, especially from local sources, might be more organic than aid sent from central bureaucracies.

Paul, a Jew, was particularly concerned about unifying Jewish and Gentile believers. Just as the early church appointed men to look after the needs of both Jewish and Hellenistic widows (Acts chapter 6), so too, Paul collected offerings from gentiles to take back to the Jewish church in Jerusalem, which was undergoing a time of need (see Acts 20:1-5; I Cor. 16:1-4; 2 Cor. 8:1-9; Romans 15: 14-32. As a missionary, however, Paul preferred to work at his own trade (tent-making) to support himself rather than relying on the charity of the churches he was visiting.

In the second passage below, taken from his second letter to the church at Thessalonica, Greece, Paul showed his concern over busybodies within the church who preferred sharing gossip over working. His admonition resonates with subsequent generations of believers, especially those from the American Puritan tradition, who attempted to distinguish the deserving from the undeserving poor and needy. This distinction often carries over into disaster relief, where governments fear that too much help from central authorities could undermine prevention efforts or local self-help during times of catastrophe.

I Corinthians 12: 25-26 (KJV)

25 That there should be no schism in the body; but that the members should have the same care one for another.

26 And whether one member suffer, all the members suffer with it;
or one member be honoured, all the members rejoice with it.

2 Thessalonians Chapter 3

6 Now we command you, brethren, in the name of our Lord Jesus
Christ, that ye withdraw yourselves from every brother that
walketh disorderly, and not after the tradition which he received
of us.
7 For yourselves know how ye ought to follow us: for we behaved
not ourselves disorderly among you;
8 Neither did we eat any man's bread for nought; but wrought with
labour and travail night and day, that we might not be chargeable
to any of you:
9 Not because we have not power, but to make ourselves an
ensample unto you to follow us.
10 For even when we were with you, this we commanded you, that
if any would not work, neither should he eat.
11 For we hear that there are some which walk among you
disorderly, working not at all, but are busybodies.
12 Now them that are such we command and exhort by our Lord
Jesus Christ, that with quietness they work, and eat their own
bread.

Fortune is the arbiter of one-half of our actions, but . . . she still leaves us to direct the other half, or perhaps a little less

---

Niccolo Machiavelli
Chapter XXV, THE PRINCE

1532

Introduction

Few political philosophers have been more notorious than Niccolo Machiavelli (1496-1527), an Italian diplomat whose best known work, *The Prince,* which was written in 1513 and published in 1532. Scholars have both praised the work for its realism and excoriated for its cynicism. In this work, Machiavelli taught that it was more important for princes to appear to be good and generous than actually to be so. He reinterpreted the idea of virtue from that of morality to that of effectiveness.

Machiavelli was an ardent patriot, who was disturbed by the wars among Italian city states, including his own native Florence. Whereas many accepted the situation as natural, Machiavelli believed that rulers could mitigate conflict by preparing for it in advance. Rather chauvinistically likening fortune to a woman who needed to be mastered, Machiavelli counseled that rather than seeking to succumb to fortune, rulers should try to control her. Perhaps not coincidentally, the ancient furies were women, and until quite recently, hurricanes were always given female names), much as one might seek to control flooding by constructing dams and dykes in advance.

Although Machiavelli was writing specifically about political preservation and preparation, his musings are relevant to the natural disasters that he used to illustrate them. His work suggests that some disasters that people attribute to "nature" or to "ill fortune" are either caused, aggravated, or could be mitigated by prior human efforts.

> To take an obvious example, if government provides flood relief by rebuilding homes in flood zones, it may lay the ground for still further catastrophes.

What Fortune Can Effect in Human Affairs, And How to Withstand Her

It is not unknown to me how many men have had, and still have, the opinion that the affairs of the world are in such wise governed by fortune and by God that men with their wisdom cannot direct them and that no one can even help them; and because of this they would have us believe that it is not necessary to labour much in affairs, but to let chance govern them. This opinion has been more credited in our times because of the great changes in affairs which have been seen, and may still be seen, every day, beyond all human conjecture. Sometimes pondering over this, I am in some degree inclined to their opinion. Nevertheless, not to extinguish our free will, I hold it to be true that Fortune is the arbiter of one-half of our actions, but that she still leaves us to direct the other half, or perhaps a little less.

I compare her to one of those raging rivers, which when in flood overflows the plains, sweeping away trees and buildings, bearing away the soil from place to place; everything flies before it, all yield to its violence, without being able in any way to withstand it; and yet, though its nature be such, it does not follow therefore that men, when the weather becomes fair, shall not make provision, both with defences and barriers, in such a manner that, rising again, the waters may pass away by canal, and their force be neither so unrestrained nor so dangerous. So it happens with fortune, who shows her power where valour has not prepared to resist her, and thither she turns her forces where she knows that barriers and defences have not been raised to constrain her.

And if you will consider Italy, which is the seat of these changes, and which has given to them their impulse, you will see it to be an open country without barriers and without any defence. For if it had been defended by proper valour, as are Germany, Spain, and France, either this invasion would not have made the great changes it has made or it would not have come at all. And this I consider enough to say concerning resistance to fortune in general.

But confining myself more to the particular, I say that a prince may be seen happy to-day and ruined to-morrow without having shown any change of disposition or character. This, I believe, arises firstly from causes that have already been discussed at length, namely, that the prince who relies entirely upon fortune is lost when it changes. I believe also that he will be successful who directs his actions according to the spirit of the times, and that he whose actions do not accord with the times will not be successful. Because men are seen, in affairs that lead to the end which every man has before him, namely, glory and riches, to get there by various methods; one with caution, another with haste; one by force, another by skill; one by patience, another by its opposite; and each one succeeds in reaching the goal by a different method. One can also see of two cautious men the one attain his end, the other fail; and similarly, two men by different observances are equally successful, the one being cautious, the other impetuous; all this arises from nothing else than whether or not they conform in their methods to the spirit of the times. This follows from what I have said, that two men working differently bring about the same effect, and of two working similarly, one attains his object and the other does not.

Changes in estate also issue from this, for if, to one who governs himself with caution and patience, times and affairs converge in such a way that his administration is successful, his fortune is made; but if times and affairs change, he is ruined if he does not change his course of action. But a man is not often found sufficiently circumspect to know how to accommodate himself to the change, both because he cannot deviate from what nature inclines him to, and also because, having always prospered by acting in one way, he cannot be persuaded that it is well to leave it; and, therefore, the cautious man, when it is time to turn adventurous, does not know how to do it, hence he is ruined; but had he changed his conduct with the times fortune would not have changed. . . .
[*Machiavelli illustrates with some examples.*]

I conclude therefore that, fortune being changeful and mankind steadfast in their ways, so long as the two are in agreement men are successful, but unsuccessful when they fall out. For my part I consider that it is better to be adventurous than cautious, because fortune is a woman, and if you wish to keep her under it is necessary to beat and ill-use her; and it is seen that she allows herself to be mastered by the adventurous rather than by those who go to work more coldly. She is, therefore, always, woman-like, a lover of young men, because they are less cautious, more violent, and with more audacity command her.

Source: http://www.constitution.org/mac/prin ce25.htm.

I am involved in mankind

John Donne
MEDITATION XVII
"FOR WHOM THE BELL TOLLS"
(NO MAN IS AN ISLAND)
1624

Introduction

One of the best known poems, which originally started as prose in Meditation 17 from his *Devotions Upon Emergent Occasions,* which was originally published in 1624, is John Donne's (1573-1631) "For Whom the Bell Tolls." Donne was an Anglican cleric who served as a member of the English Parliament from 1601 to 1614. He is buried at St. Paul's Cathedral in London.

This poem stresses the interconnectedness of humanity and mighty therefore provide justification for both domestic and international disaster aid. Interestingly, Donne connects the fate of his native island to that of Europe and the rest of mankind.

No man is an island,
Entire of itself,
Every man is a piece of the continent,
A part of the main.
If a clod be washed away by the sea,
Europe is the less.
As well as if a promontory were.
As well as if a manor of thy friend's
Or of thine own were:
Any man's death diminishes me,
Because I am involved in mankind,
And therefore never send to know for whom the bell tolls;
It tolls for thee.
John Donne

[S]uch a war as is of every man against every man

Thomas Hobbes
LEVIATHAN
1651

Introduction

Early American thought borrowed heavily from social contract thinking, as chiefly articulated by Thomas Hobbes (1588-1679) and John Locke (1632-1704). Both sought to explain the role of government by hypothesizing about what the situation would be without it. In *Leviathan,* Hobbes painted an especially grim portrait, which he likened to a war of all against all, in which the life of man would be "solitary, poor, nasty, brutish, and short." Locke's portrayal in his *Second Treatise on Government,* was not quite so dour, but he anticipated that in a stateless state, laws would not be established, settled, or known; laws would have no arbiter (such as modern judges); and laws would have no enforcer (like modern executives). In such a situation, individuals' lives, liberties, and property (what Thomas Jefferson would later call "the pursuit of happiness") would be jeopardized. Individuals would accordingly attempt to form a "social contract" to secure and protect their rights.

Naomi Zack (2006) has argued that in disasters, individuals often enter into a second state of nature where societal institutions break down, and many individuals often have to resort to self help. Although stories about disasters often emphasize those who are looting or otherwise taking advantage of the situation, however, humans often act altruistically in such circumstances, seeking to rescue and aid fellow individuals even before governmental aid arrives.

THE STATE OF NATURE

Chapter XIII

Of the Natural Condition of Mankind as Concerning Their Felicity and Misery

NATURE hath made men so equal in the faculties of the body and mind, as that, though there be found one man sometimes manifestly stronger in body or of quicker mind than another, yet when all is reckoned together the difference between man and man is not so considerable as that one man can thereupon claim to himself any benefit to which another may not pretend as well as he. For, as to the strength of body, the weakest has strength enough to kill the strongest, either by secret machination or by confederacy with others that are in the same danger with himself.

And, as to the faculties of the mind, setting aside the arts grounded upon words and especially that skill of proceeding upon general and infallible rules called science, which very few have and but in few things, as being not a native faculty born with us, nor attained, as prudence, while we look after somewhat else, I find yet a greater equality amongst men than that of strength. For prudence is but experience, which equal time equally bestows on all men in those things they equally apply themselves unto. That which may perhaps make such equality incredible is but a vain conceit of one's own wisdom, which almost all men think they have in a greater degree than the vulgar, that is, than all men but themselves, and a few others whom by fame or for concurring with themselves they approve. For such is the nature of men that, howsoever they may acknowledge many others to be more witty or more eloquent or more learned, yet they will hardly believe there be many so wise as themselves, for they see their own wit at hand and other men's at a distance. But this proveth rather that men are in that point equal than unequal. For there is not ordinarily a greater sign of the equal distribution of anything than that every man is contented with his share.

From this equality of ability ariseth equality of hope in the attaining of our ends. And therefore, if any two men desire the same thing which nevertheless they cannot both enjoy, they become enemies; and, in the way to their end, which is principally their own conservation and sometimes their delectation only, endeavour to destroy or subdue one another. And from hence it comes to pass that, where an invader hath no more to fear than another man's single power, if one plant, sow, build, or possess, a convenient seat others may probably be expected to come prepared with forces united to dispossess and deprive him not only of the fruit of his labor but also of his life or liberty. And the invader again is in the like danger of another.

And from this diffidence of one another there is no way for any man to secure himself so reasonable as anticipation, that is, by force or wiles to master the persons of all men he can so long till he see no other power great enough to endanger him; and this is no more than his own conservation requireth and is generally allowed. Also, because there be some that, taking pleasure in contemplating their own power in the acts of conquest, which they pursue farther than their security requires, if others, that otherwise would be glad to be at ease within the modest bounds, should not be invasion increase their power, they would not be able long time, by standing only on their defence, to subsist. And by consequence, such augmentation of dominion over men being necessary to a man's conservation, it ought to be allowed him.

Again, men have no pleasure, but on the contrary a great deal of grief, in keeping company where there is no power able to overawe them all. For every man looketh that his companion should value him at the same rate he sets upon himself, and, upon all signs of contempt or undervaluing, naturally endeavours as far as he dares (which amongst them that have no common power to keep them in quiet, is far enough to make them destroy each other) to extort a greater value from his contemners by damage, and from others by the example.

So that in the nature of man we find three principal causes of quarrel. First, competition; secondly, diffidence; thirdly, glory.

The first maketh man invade for gain; the second, for safety; and the third, for reputation. The first use violence, to make themselves masters of other men's persons, wives, children, and cattle; the second, to defend them; the third, for trifles, as a word, a smile, a different opinion, and any other sign of undervalue, either direct in their persons or by reflection in their kindred, their friends, their nation, their profession, or their name

Hereby it is manifest that, during the time men live without a common power to keep them all in awe, they are in that condition which is called war, and such a war as is of every man against every man. For 'war' consisteth not in battle only or the act of fighting, but in a tract of time wherein the will to contend by battle is sufficiently known, and therefore the notion of 'time' is to be considered in the nature of war, as it is in the nature of weather. For as the nature of foul weather lieth not in a shower or two of rain but in an inclination thereto of many days together, so the nature of war consisteth not in actual fighting but in the known disposition thereto during all the time there is no assurance to the contrary. All other time is 'peace.'

Whatsoever therefore is consequent to a time or war where every man is enemy to every man, the same is consequent to the time wherein men live without other security than what their own strength and their own invention shall furnish them withal. In such condition there is no place for industry, because the fruit thereof is uncertain, and consequently no culture of the earth, no navigation nor use of the commodities that may be imported by sea, no commodious building, no instruments of moving and removing such things as require much force, no knowledge of the face of the earth; no account of time, no arts, no letters, no society, and, which is worst of all, continual fear and danger of violent death, and the life of man solitary, poor, nasty, brutish, and short.

It may seem strange to some man that has not well weighed these things that Nature should thus dissociate and render men apt to invade and destroy one another; and he may therefore, not trusting to this inference made from the passions, desire perhaps to have the same confirmed by experience. Let him therefore consider with himself, when taking a journey, he arms himself and seeks to go well accompanied; when going to sleep, he locks his doors; when even in his house, he locks his chests; and this when he knows there be laws and public officers armed to revenge all injuries shall be done him; what opinion he has of his fellow-subjects when he rides armed; of his fellow-citizens, when he locks his doors; and of his children and servants, when he locks his chests. Does he not there as much accuse mankind by his actions as I do by my words? But neither of us accuse man's nature in it. The desires and other passions of man are in themselves no sin. No more are the actions that proceed from those passions, till they know a law that forbids them; which, till laws be made, they cannot know, nor can any law be made till they have agreed upon the person that shall make it.

It may peradventure be thought there was never such a time nor condition of war as this; and I believe it was never generally so over all the world, but there are many places where they live so now. For the savage people in many places of America, except the government of small families the concord whereof dependeth on natural lust, have no government at all, and live at this day in that brutish manner as I said before. Howsoever, it may be perceived what manner of life there would be where there were no common power to fear, by the manner of life which men that have formerly lived under a peaceful government use to degenerate into, in a civil war.

But, though there had never been any time wherein particular men were in a condition of war one against another, yet in all times kings and persons of sovereign authority, because of their independency, are in continual jealousies and in the state and posture of gladiators, having their weapons pointing, and their eyes fixed on one another, that is, their forts, garrisons, and guns, upon the frontiers of their kingdoms, and continual spies upon their neighbours: which is a posture of war. But because they uphold thereby the industry of their subjects, there does not follow from it that misery which accompanies the liberty of particular men.

To this war of every man against every man this also is consequent, that nothing can be unjust. The notions of right and wrong, justice and injustice, have there no place. Where there is no common power, there is no law; where no law, no injustice. Force and fraud are in war the two cardinal virtues. Justice and injustice are none of the faculties neither of the body nor mind. If they were, they might be in a man that were alone in the world, as well as his senses and passions. They are qualities that relate to men in society, not in solitude. It is consequent also to the same condition that there be no propriety, no dominion, no 'mine' and 'thine' distinct, but only that to be every man's that he can get, and for so long as he can keep it. And thus much for the ill condition which man by mere nature is actually placed in, though with a possibility to come out of it, consisting partly in the passions, partly in his reason.

The passions that incline men to peace are fear of death, desire of such things as are necessary to commodious living, and a hope by their industry to obtain them. And reason suggesteth convenient articles of peace, upon which men may be drawn to agreement. These articles are they which otherwise are called the Laws of Nature, whereof I shall speak more particularly in the two following chapters.

Source: Bartleby.com. Great Books Online. http://www.bartleby.com/34/5/13.html. The Harvard Classics, 1909-14.

[T]hose succours, those duties,
which men owe to each other, as men

Emerich De Vattel
THE LAW OF NATIONS
(1758)

Introduction

Although the primary focus of this book is on aid that one part of the nation extends to another, the U.S. also devotes considerable resources to foreign aid. Although much of this is military, the government also directs some aid to humanitarian disasters like famines, earthquakes, tsunamis, disease, and the results of war. Lack of such provisions may increase international migration.

In his classic, *The Law of Nations*, which was first published in 1758, Emerich de Vattel, a Swiss philosopher (1714-1767) grappled with the issue of when it was appropriate for nations to provide aid to others, but many of his arguments would also serve as arguments for governments to aid their own citizens. Vattel seems to recognize that nations should not impose unwanted aid on nations who do not want it, but this sometimes means that outside nations are prevented from helping those in need by the callousness, and fears, of their own governments.

BOOK II
OF A NATION CONSIDERED IN ITS RELATION TO OTHERS
CHAP. I.
OF THE COMMON DUTIES OF A NATION TOWARDS OTHERS; OR, OF THE OFFICES OF HUMANITY BETWEEN NATIONS.

§ 1. Foundation of the common and mutual duties of nations.

THE following maxims will appear very strange to cabinet politicians; and such is the misfortune of mankind, that, to many of those refined conductors of nations, the doctrine of this chapter will be a subject of ridicule. Be it so; but we will, nevertheless, boldly lay down what the law of nature prescribes to nations. . . .

And why should we not hope still to find, among those who are the head of affairs, come wise individuals who are convinced of this great truth, that virtue is, even for sovereigns and political bodies, the most certain road to prosperity and happiness? There is at least one benefit to be expected from the open assertion and publication of sound maxims, which is, that even those who relish them the least are thereby laid under a necessity of keeping within some bounds, lest they should forfeit their characters altogether. To flatter ourselves with the vain expectation that men, and especially men in power, will be inclined strictly to conform to the laws of nature, would be a gross mistake; and to renounce all hope of making impression on some of them, would be to give up mankind for lost.

Nations, being obliged by nature reciprocally to cultivate human society (Prelim. § 11), are bound to observe towards each other all the duties which the safety and advantage of that society require.

§ 2. Offices of humanity, and their foundation.

The offices of humanity are those succours, those duties, which men owe to each other, as men, — that is, as social beings formed to live in society, and standing in need of mutual assistance for their preservation and happiness, and to enable them to live in a manner conformable to their nature. Now, the laws of nature being no less obligatory on nations than on individuals (Prelim. § 5), whatever duties each man owes to other men, the same does each nation, in its way, owe to other nations (Prelim. § 10, &c). Such is the foundation of those common duties — of those offices of humanity — to which nations are reciprocally bound towards each other. They consist, generally, in doing every thing in our power for the preservation and happiness of others, as far as such conduct is reconcilable with our duties towards ourselves.

§ 3. General principle of all the mutual duties of nations.

The nature and essence of man, who, without the assistance of his fellow-men, is unable to supply all his wants, to preserve himself, to render himself perfect, and to live happily, plainly show us that he is destined to live in society, in the interchange of mutual aid; and, consequently, that all men are, by their very nature and essence, obliged to unite their common efforts for the perfection of their own being and that of their condition. The surest method of succeeding in this pursuit is, that each individual should exert his efforts first for himself and then for others. Hence it follows, that, whatever we owe to ourselves, we likewise owe to others, so far as they stand in need of assistance, and we can grant it to them without being wanting to ourselves. Since, then, one nation, in its way, owes to another nation every duty that one man owes to another man, we may confidently lay down this general principle: — one state owes to another state whatever it owes to itself, so far as that other stands in real need of its assistance, and the former can grant it without neglecting the duties it owes to itself. Such is the eternal and immutable law of nature. Those who might be alarmed at this doctrine, as totally subversive of the maxims of sound policy, will be relieved from their apprehensions by the two following considerations:

1. Social bodies or sovereign states are much more capable of supplying all their wants than individual men are; and mutual assistance is not so necessary among them, nor so frequently required. Now, in those particulars which a nation can itself perform, no succour is due to it from others.

2. The duties of a nation towards itself, and chiefly the care of its own safety, require much more circumspection and reserve than

need be observed by an individual in giving assistance to others. This remark we shall soon illustrate.

§ 4. Duties of a nation for the preservation of others.

Of all the duties of a nation towards itself, the chief object is its preservation and perfection, together with that of its state. The detail given of them in the first book of this work may serve to point out the several objects in relation to which a state may and should assist another state. Every nation ought, on occasion, to labour for the preservation of others, and for securing them from ruin and destruction, as far as it can do this without exposing itself too much. Thus, when a neighbouring nation is unjustly attacked by a powerful enemy who threatens to oppress it, if you can defend it, without exposing yourself to great danger, unquestionably it is your duty to do so. Let it not be said, in objection to this, that a sovereign is not to expose the lives of his soldiers for the safety of a foreign nation with which he has not contracted a defensive alliance. It may be his own case to stand in need of assistance; and, consequently, he is acting for the safety of his own nation in giving energy to the spirit and disposition to afford mutual aid. Accordingly, policy here coincides with and enforces obligation and duty. It is the interest of princes to stop the progress of an ambitious monarch, who aims at aggrandizing himself by subjugating his neighbours. A powerful league was formed in favour of the United Provinces, when threatened with the yoke of Louis XIV.[3] When the Turks laid siege to Vienna, the brave Sobieski, king of Poland, saved the house of Austria and possibly all Germany, and his own kingdom.

§ 5. It ought to assist a nation afflicted with famine or any other calamities.

For the same reason, if a nation is afflicted with famine, all those who have provisions to spare ought to relieve her distress, without, however, exposing themselves to want. But, if that nation is able to pay for the provisions thus furnished, it is perfectly lawful to sell them to her at a reasonable rate; for they are not bound to furnish her with what she is herself capable of procuring; and, consequently, there is no obligation of gratuitously bestowing on her such things as she is able to purchase. To give assistance in such extreme necessity is so essentially conformable to humanity, that the duty is seldom neglected by any nation that has received the slightest polish of civilization. The great Henry the Fourth could not forbear to comply with it in favour of obstinate rebels who were bent on his destruction.

Whatever be the calamity with which a nation is afflicted, the like assistance is due to it. We have seen little states in Switzerland order public collections to be made in behalf of towns or villages of the neighbouring countries, which had been ruined by fire, and remit them liberal succours; the difference of religion proving no bar to the performance of so humane a deed. The calamities of Portugal have given England an opportunity of fulfilling the duties of humanity with that noble generosity which characterizes a great nation. On the first intelligence of the disastrous fate of Lisbon, the parliament voted a hundred thousand pounds sterling for the relief of an unfortunate people; the king also added considerable sums: ships, laden with provisions and all kinds of succours, were sent away with the utmost dispatch; and their arrival convinced the Portuguese that an opposition in belief and worship does not restrain the beneficence of those who understand the claims of humanity. On the same occasion, likewise, the king of Spain signally displayed his tenderness for a near ally, and exerted, in a conspicuous manner, his humanity and generosity.

§ 6. It ought to contribute to the perfection of other states.

A nation must not simply confine itself to the preservation of other states; it should likewise, according to its power and their want of its assistance, contribute to their perfection. We have already shown (Prelim. § 13) that natural society imposes on it this general obligation. We are now come to the proper place for treating of the obligation somewhat more in detail. A state is more or less perfect, as it is more or less adapted to attain the end of civil society, which consists in procuring for its members every thing of which they stand in need, for the necessities, the conveniences, and enjoyments of life, and for their happiness in general, — in providing for the peaceable enjoyment of property, and the safe and easy administration of justice, — and, finally, in defending itself against all foreign violence (Book I. § 15). Every nation therefore, should occasionally, and according to its power, contribute, not only to put another nation in possession of these advantages, but likewise to render it capable of procuring them itself. Accordingly, a learned nation, if applied to for masters and teachers in the sciences, by another nation desirous of shaking off it native barbarism, ought not to refuse such a request. A nation, whose happiness it is to live under wise laws, should on occasion, make it a point of duty to communicate them. Thus, when the wise and virtuous Romans sent ambassadors to Greece to collect good laws, the Greeks were far from rejecting so reasonable and so laudable a request.

§ 7. But not by force.

But, though a nation be obliged to promote, as far as lies in its power, the perfection of others, it is not entitled forcibly to obtrude these good offices on them. Such an attempt would be a violation of their natural liberty. In order to compel any one to receive a kindness, we must have an authority over him; but nations are absolutely free and independent (Prelim. § 4). Those ambitious Europeans who attacked the American nations, and subjected them to their greedy dominion, in order, as they pretended, to civilize them, and cause them to be instructed in the true religion, — those usurpers, I say, grounded themselves on a pretext equally unjust and ridiculous. It is strange to hear the learned and judicious Grotius assert that a sovereign may justly take up arms to chastise nations which are guilty of enormous transgressions of the law of nature, which treat their parents with inhumanity like the Sogdians, which eat human flesh as the ancient Gauls,& c.) What led him into this error, was, his attributing to every independent man, and of course to every sovereign, an odd kind of right to punish faults which involve an enormous violation of the laws of nature, though they do not affect either his rights or his safety. But we have shown (Book I. § 169) that men derive the right of punishment solely from their right to provide for their own safety; and consequently they cannot claim it except against those by whom they have been injured. Could it escape Grotius, that, notwithstanding all the precautions added by him in the following paragraphs, his opinion opens a door to all the ravages of enthusiasm and fanaticism, and furnishes ambition with numberless pretexts? Mohammed and his successors have desolated and subdued Asia, to avenge the indignity done to the unity of the Godhead; all whom they termed associators or idolaters fell victims to their devout fury.

§ 8. The right to require the offices of humanity.

Since nations ought to perform these duties or offices of humanity towards each other, according as one stands in need, and the other can reasonably comply with them, — every nation being free, independent, and sole arbitress of her own actions, it belongs to each to consider whether her situation warrants her in asking or granting any thing on this head. Thus 1. Every nation has a perfect right to ask of another that assistance and those kind offices which she conceives herself to stand in need of. To prevent her, would be doing her an injury. If she makes the application without necessity, she is guilty of a breach of duty; but, in this respect, she is wholly independent of the judgment of others. A nation has a right to ask for these kind offices, but not to demand them.

§ 9. The right of judging whether they are to be granted.

For, 2. These offices being due only in necessity, and by a nation which can comply with them without being wanting to itself; the nation that is applied to has, on the other hand, a right of judging whether the case really demands them, and whether circumstances will allow her to grant them consistently with that regard which she ought to pay to her own safety and interests: for instance, a nation is in want of corn, and applies to another nation to sell her a quantity of it: — in this case it rests with the latter party to judge whether, by a compliance with the request, they will not expose themselves to the danger of a scarcity: and, if they refuse to comply, their determination is to be patienty acquiesced in. We have very lately seen a prudent performance of this duty on the part of Russia: she generously assisted Sweden when threatened with a famine, but refused to other powers the liberty of purchasing corn in Livonia, from the circumstance of standing herself in need of it, and, no doubt, from weighty political motives likewise.

§ 10. A nation is not to compel another to perform these.

Thus, the right which a nation has to the offices of humanity is but an imperfect one: she cannot compel another nation to the performance of them. The nation that unreasonably refuses them offends against equity, which consists in acting conformably to the imperfect right of another: but thereby no injury is done; injury or injustice being a trespass against the perfect right of another.

§ 11. Mutual love of nations.

It is impossible that nations should mutually discharge all these several duties if they do not love each other. This is the pure source from which the offices of humanity should proceed; they will retain the character and perfection of it. Then nations will be seen sincerely and cheerfully to help each other, earnestly to promote their common welfare, and cultivate peace, without jealousy or distrust.

§ 12. Each nation ought to cultivate the friendship of others.

A real friendship will be seen to reign among them; and this happy state consists in a mutual affection, Every nation is obliged to cultivate the friendship of other nations, and carefully to avoid whatever might kindle their enmity against her. Wise and prudent nations often pursue this line of conduct from views of direct and present interest: a more noble, more general, and less direct interest, is too rarely the motive of politicians. If it be incontestable that men must love each other in order to answer the views of nature and discharge the duties which she prescribes them, as well as for their own private advantage, — can it be doubted that nations are under the like reciprocal obligation? Is it in the power of men, on dividing themselves into different political bodies, to break the ties of that universal society which nature has established amongst them?

§ 13. To perfect itself with a view to the advantage of others, and set them good examples.

If a man ought to qualify himself for becoming useful to other men, — and a citizen, for rendering useful services to his country and fellow citizens, a nation likewise, in perfecting herself, ought to have in view the acquisition of a greater degree of ability to promote the perfection and happiness of other nations; she should be careful to set them good examples, and avoid setting them a pattern of any thing evil. Imitation is natural to mankind: the virtues of a celebrated nation are sometimes imitated, and much more frequently its vices and defects.

§ 14. To take care of their glory.

Glory being a possession of great importance to a nation, as we have shown in a particular chapter expressly devoted to the subject,

— the duty of a nation extends even to the care of the glory of other nations. In the first place, she should, on occasion, contribute to enable them to merit true glory: secondly, she should do them in this respect all the justice due to them, and use all proper endeavours that such justice be universally done them: finally, instead of irritating, she should kindly extenuate the bad effect which some slight blemishes may produce.

§ 15. Difference of religion.

From the manner in which we have established the obligation of performing the offices of humanity, it plainly appears to be solely founded on the nature of man. Wherefore, no nation can refuse them to another, under pretence of its professing a different religion; to be entitled to them, it is sufficient that the claimant is our fellow-creature, A conformity of belief and worship may become a new tie of friendship between nations: but no difference in these respects can warrant us in laying aside the character of men, or the sentiments annexed to it. As we have already related (§ 5) some instances well worthy of imitation, let us here do justice to the pontiff who at present fills the see of Rome, and has recently given a very remarkable example, and which cannot be loo highly commended. Information being given to that prince, that several Dutch ships remained at Civita Vecchia, not daring to put to sea for fear of the Algerine corsairs, he immediately issued orders that the frigates of the ecclesiastical state should convoy those ships out of danger; and his nuncio at Brussels received instructions to signify to the ministers of the states-general, that his holiness made it a rule to protect commerce and perform the duties of humanity, without regarding any difference of religion. Such exalted sentiments cannot fail of raising a veneration for Benedict XIV even amongst Protestants.

§ 16. Rule and measure of the offices of humanity.

How happy would mankind be, were these amiable precepts of nature everywhere observed! Nations would communicate to each other their products and their knowledge; a profound peace would prevail all over the earth, and enrich it with its invaluable fruits; industry, the sciences and the arts would be employed in promoting our happiness, no less than in relieving our wants; violent methods of deciding contests would be no more heard of; all differences would be terminated by moderation, justice, and equity; the world would have the appearance of a large republic; men would live everywhere like brothers, and each individual be a citizen of the universe. That this idea should be but a delightful dream! yet it flows from the nature and essence of man. Put disorderly passions, and private and mistaken interest, will for ever prevent its being realized. Let us then, consider what limitations the present state of men, and the ordinary maxims and conduct of nations, may render necessary in the practice of these precepts of nature, which are in themselves so noble and excellent.

The law of nature cannot condemn the good to become the dupes and prey of the wicked, and the victims of their injustice and ingratitude. Melancholy experience shows that most nations aim only to strengthen and enrich themselves at the expense of others, — to domineer over them, and even if an opportunity offers, to oppress and bring them under the yoke. Prudence does not allow us to strengthen an enemy or one in whom we discover a desire of plundering and oppressing us: and the care of our own safety forbids it. We have seen (§ 3, &c.) that a nation does not owe her assistance and the offices of humanity to other nations, except so far as the grant of them is reconcilable with her duties to herself. Hence, it evidently follows, that, though the universal love of mankind obliges us to grant at all times, and to all, even to our enemies, those offices which can only tend to render them more moderate and virtuous, because no inconvenience is to be apprehended from granting them, — we are not obliged to give them such succours as probably may become destructive to ourselves. Thus, 1. The exceeding importance of trade, not only to the wants and conveniences of life, but likewise to the strength of a state, and furnishing it with the means of defending itself against its enemies, — and the insatiable avidity of those nations which seek wholly and exclusively to engross it, — thus, I say, these circumstances authorize a nation possessed of a branch of trade, or the secret of some important manufacture or fabric, to reserve to herself those sources of wealth, and, instead of communicating them to foreign nations, to take measure against it. But, where the necessaries or conveniences of life are in question, the nation ought to sell them to others at a reasonable price, and not convert her monopoly into a system of odious extortion. To commerce England chiefly owes her greatness, her power, and her safety: who, then, will presume to blame her for endeavouring, by every fair and just method, to retain the several branches of it in her own hand?

2. As to things directly and more particularly useful for war, a nation is under no obligation to sell them to others of whom it has the smallest suspicion; and prudence even declares against it. Thus, by the Roman laws, people were very justly prohibited to instruct the barbarous nations in building galleys. Thus, in England, laws have been enacted to prevent the best method of ship-building from being carried out of the kingdom.

This caution is to be carried farther, with respect to nations more justly suspected. Thus, when the Turks were successfully pursuing their victorious career, and rapidly advancing to the zenith of power, all Christian nations ought, independent of every bigoted consideration, to have considered them as enemies; even the most distant of those nations, though not engaged in any contest with them, would have been justifiable in breaking off all commerce with a people who made it their profession to subdue by force of arms all who would not acknowledge the authority of their prophet.

§ 17. Particular limitation with regard to the prince.

Let us further observe, with regard to the prince in particular, that he ought not, in affairs of this nature, to obey without reserve all the suggestions of a noble and generous heart impelling him to sacrifice his own interests to the advantage of others, or to motives of generosity; because it is not his private interest that is in question, but that of the state — that of the nation who has committed herself to his care. Cicero says that a great and elevated soul despises pleasures, wealth, life itself, and makes no account of them, when the common utility is at stake. He is right, and such sentiments are to be admired in a private person; but generosity is not to be exerted at the expense of others. The head or conductor of a nation ought not to practise that virtue in public affairs without great circumspection, nor to a greater extent than will redound to the glory and real advantage of the state. As to the common good of human society, he ought to pay the same attention to it as the nation he represents would be obliged to pay were the government of her affairs in her own hand.

§ 18. No nation ought to injure others.

But, though the duties of a nation towards herself set bounds to the obligation of performing the offices of humanity, they cannot in the least affect the prohibition of doing any harm to others, of causing them any prejudice, — in a word, of injuring them ... If every man is, by his very nature, obliged to assist in promoting the perfection of others, much more cogent are the reasons which forbid him to increase their imperfection, and that of their condition. The same duties are incumbent on nations (Prelim. §§ 5, 6). No nation, therefore, ought to commit any actions tending to impair the perfection of other nations, and that of their condition, or to impede their progress, — in other words, to injure them. And, since the perfection of a nation consists in her aptitude to attain the end of civil society — and the perfection of her condition, in not wanting any of the things necessary to that end (Book I. § 14) — no one nation ought to hinder another from attaining the end of civil society, or to render her incapable of attaining it. This general principle forbids nations to practise any evil manœuvres tending to create disturbance in another state, to foment discord, to corrupt its citizens, to alienate its allies, to raise enemies against it, to tarnish its glory, and to deprive it of its natural advantages.

However, it will be easily conceived that negligence in fulfilling the common duties of humanity, and even the refusal of these duties or offices, is not an injury. To neglect or refuse contributing to the perfection of a nation, is not impairing that perfection.

It must be further observed, that, when we are making use of our right, when we are doing what we owe to ourselves or to others, if, from this action of ours, any prejudice results to the perfection of another, — any detriment to his exterior condition, — we are not guilty of an injury we are doing what is lawful, or even what we ought to do. The damage which accrues to the other is no part of our intention: it is merely an accident, the imputability of which must be determined by the particular circumstances. For instance, in case of a lawful defence, the harm we do to the aggressor is not the object we aim at; — we act only with a view to our own safety;

we make use of our right; and the aggressor alone is chargeable with the mischief which he brings on himself.

§ 19. Offences.

Nothing is more opposite to the duties of humanity, nor more contrary to that society which should be cultivated by nations, than offences, or actions which give a just displeasure to others: every nation therefore should carefully avoid giving any other nation real offence: I say real; for, should others take offence at our behaviour when we are only using our rights or fulfilling our duties, the fault lies with them, not with us. Offences excite such asperity and rancour between nations that we should avoid giving any room even for ill-grounded piques, when it can be done without any inconveniency, or failure in our duty. It is said that certain medals and dull jests irritated Louis XIV. against the United Provinces to such a degree as to induce him, in 1672, to undertake the destruction of that republic. (96)

§ 20. Bad customs of the ancients.

The maxims laid down in this chapter, — those sacred precepts of nature, — were for a long time unknown to nations. The ancients had no notion of any duty they owed to nations with whom they were not united by treaties of friendship. The Jews especially placed a great part of their zeal in hating all nations; and, as a natural consequence, they were detested and despised by them in turn. At length the voice of nature came to be heard among civilized nations; they perceived that all men are brethren. When will the happy time come that they shall behave as such?

Source: http://www.constitution.org/vattel/vattel_02.htm. Accessed 12/6/2016.

Things could not be otherwise than they were

Voltaire
CANDIDE
1759

Introduction

Natural disasters raise important philosophical questions about the nature of the universe. In his satirical novel, *Candidem* (1759), Voltaire, the pen name for Francois-Marie Aroet, a French philosopher who lived from 1694-1778, used natural disasters (especially the Lisbon, Portugal earthquake of 1755) to call into question Gottfried Leibniz's (1646-1716) optimistic philosophical view, represented in this narrative by Dr. Pangloss, that the world was the best of all possible worlds. As Dr. Pangloss suggests, such a view (not unlike modern arguments for fatalism or for "survival of the fittest") might caution against human intervention. If, by contrast, people recognized natural disasters as true evils that might be mitigated, then human intervention is warranted.

Chapter 5
A Tempest, a Shipwreck, an Earthquake, and What Else Befell Dr. Pangloss, Candide, and James, the Anabaptist

One half of the passengers, weakened and half-dead with the inconceivable anxiety and sickness which the rolling of a vessel at sea occasions through the whole human frame, were lost to all sense of the danger that surrounded them. The others made loud outcries, or betook themselves to their prayers; the sails were blown into shreds, and the masts were brought by the board. The vessel was a total wreck. Everyone was busily employed, but nobody could be either heard or obeyed. The Anabaptist, being upon deck, lent a helping hand as well as the rest, when a brutish sailor gave him a blow and laid him speechless; but, not withstanding, with the violence of the blow the tar himself tumbled headforemost overboard, and fell upon a piece of the broken mast, which he immediately grasped.

Honest James, forgetting the injury he had so lately received from him, flew to his assistance, and, with great difficulty, hauled him in again, but, not withstanding, in the attempt, was, by a sudden jerk of the ship, thrown overboard himself, in sight of the very fellow whom he had risked his life to save and who took not the least notice of him in this distress. Candide, who beheld all that passed and saw his benefactor one moment rising above water, and the next swallowed up by the merciless waves, was preparing to jump after him, but was prevented by the philosopher Pangloss, who demonstrated to him that the roadstead of Lisbon had been made on purpose for the Anabaptist to be drowned there. While he was proving his argument a priori, the ship foundered, and the whole crew perished, except Pangloss, Candide, and the sailor who had been the means of drowning the good Anabaptist. The villain swam ashore; but Pangloss and Candide reached the land upon a plank.

As soon as they had recovered from their surprise and fatigue they walked towards Lisbon; with what little money they had left they thought to save themselves from starving after having escaped drowning.

Scarcely had they ceased to lament the loss of their benefactor and set foot in the city, when they perceived that the earth trembled under their feet, and the sea, swelling and foaming in the harbor, was dashing in pieces the vessels that were riding at anchor. Large sheets of flames and cinders covered the streets and public places; the houses tottered, and were tumbled topsy-turvy even to their foundations, which were themselves destroyed, and thirty thousand inhabitants of both sexes, young and old, were buried beneath the ruins.

The sailor, whistling and swearing, cried, "Damn it, there's something to be got here."

"What can be the sufficing reason of this phenomenon?" said Pangloss.

"It is certainly the day of judgment," said Candide.

The sailor, defying death in the pursuit of plunder, rushed into the midst of the ruin, where he found some money, with which he got drunk, and, after he had slept himself sober he purchased the favors of the first good-natured wench that came in his way, amidst the ruins of demolished houses and the groans of half-buried and expiring persons.

Pangloss pulled him by the sleeve. "Friend," said he, "this is not right, you trespass against the universal reason, and have mistaken your time."

"Death and zounds!" answered the other, "I am a sailor and was born at Batavia, and have trampled four times upon the crucifix in as many voyages to Japan; you have come to a good hand with your universal reason."

In the meantime, Candide, who had been wounded by some pieces of stone that fell from the houses, lay stretched in the street, almost covered with rubbish.

"For God's sake," said he to Pangloss, "get me a little wine and oil! I am dying."

"This concussion of the earth is no new thing," said Pangloss, "the city of Lima in South America experienced the same last year; the same cause, the same effects; there is certainly a train of sulphur all the way underground from Lima to Lisbon."

"Nothing is more probable," said Candide; "but for the love of God a little oil and wine."

"Probable!" replied the philosopher, "I maintain that the thing is demonstrable."

Candide fainted away, and Pangloss fetched him some water from a neighboring spring. The next day, in searching among the ruins, they found some eatables with which they repaired their exhausted strength. After this they assisted the inhabitants in relieving the distressed and wounded. Some, whom they had humanely assisted, gave them as good a dinner as could be expected under such terrible circumstances. The repast, indeed, was mournful, and the company moistened their bread with their tears; but Pangloss endeavored to comfort them under this affliction by affirming that things could not be otherwise that they were.

"For," said he, "all this is for the very best end, for if there is a volcano at Lisbon it could be in no other spot; and it is impossible but things should be as they are, for everything is for the best."

By the side of the preceptor sat a little man dressed in black, who was one of the familiars of the Inquisition. This person, taking him up with great complaisance, said, "Possibly, my good sir, you do not believe in original sin; for, if everything is best, there could have been no such thing as the fall or punishment of man."

Your Excellency will pardon me," answered Pangloss, still more politely; "for the fall of man and the curse consequent thereupon necessarily entered into the system of the best of worlds."

"That is as much as to say, sir," rejoined the familiar, "you do not believe in free will."

"Your Excellency will be so good as to excuse me," said Pangloss, "free will is consistent with absolute necessity; for it was necessary we should be free, for in that the will-"

Pangloss was in the midst of his proposition, when the familiar beckoned to his attendant to help him to a glass of port wine.

Source: Online Literature Library, sponsored by KnowledgeMattersLtd.

If he was to lose his little finger to-morrow, he would not sleep tonight; but, provided he never saw them, he will snore with the most profound security over the ruin of a hundred millions of his brethren

---

Adam Smith
THE THEORY OF MORAL SENTIMENTS
1759

Introduction

Although Adam Smith, the author of *The Wealth of Nations* (1776), is the primary originator of laissez-faire economics and the idea that the "invisible hand" of the market will solve many problems of distribution, in this selection from his *Theory of Moral Sentiments*, first published in 1759, Smith makes it clear that he recognizes the manner in which their consciences enable humans to sympathize with the needs of others. His illustration, comparing an individual's concern over a finger with his relative unconcern over the fate of a distant people, is undoubtedly intended to shock the conscience of his readers. It may also explain how the rise of mass media has directed increased attention to, and motivated aid for, disasters that might otherwise be ignored. Smith's narrative does not articulate the degree to which government should assume charitable roles.

Part III: Of the Foundation of our Judgments concerning our own Sentiments and Conduct, and of the Sense of Duty Consisting of One Section
Chapter III: Of the Influences and Authority of Conscience

But though the approbation of his own conscience can scarce, upon some extraordinary occasions, content the weakness of man; though the testimony of the supposed impartial spectator, of the great inmate of the breast, cannot always alone support him; yet the influence and authority of this principle is, upon all occasions, very great; and it is only by consulting this judge within, that we can ever see what relates to ourselves in its proper shape and dimensions; or that we can ever make any proper comparison between our own interests and those of other people.

As to the eye of the body, objects appear great or small, not so much according to their real dimensions, as according to the nearness or distance of their situation; so do they likewise to what may be called the natural eye of the mind: and we remedy the defects of both these organs pretty much in the same manner. In my present situation an immense landscape of lawns, and woods, and distant mountains, seems to do no more than cover the little window which I write by and to be out of all proportion less than the chamber in which I am sitting. I can form a just comparison between those great objects and the little objects around me, in no other way, than by transporting myself, at least in fancy, to a different station, from whence I can survey both at nearly equal distances, and thereby form some judgment of their real proportions. Habit and experience have taught me to do this so easily and so readily, that I am scarce sensible that I do it; and a man must be, in some measure, acquainted with the philosophy of vision, before he can be thoroughly convinced, how little those distant objects would appear to the eye, if the imagination, from a knowledge of their real magnitudes, did not swell and dilate them.

In the same manner, to the selfish and original passions of human nature, the loss or gain of a very small interest of our own, appears to be of vastly more importance, excites a much more passionate joy or sorrow, a much more ardent desire or aversion, than the greatest concern of another with whom we have no particular connexion. His interests, as long as they are surveyed from this station, can never be put into the balance with our own, can never restrain us from doing. whatever may tend to promote our own, how ruinous soever to him. Before we can make any proper comparison of those opposite interests, we must change our position. We must view them, neither from our own place nor yet from his, neither with our own eyes nor yet with his, but from the place and with the eyes of a third person, who has no particular connexion with either, and who judges with impartiality between us. Here, too, habit and experience have taught us to do this so easily and so readily, that we are scarce sensible that we do it; and it requires, in this case too, some degree of reflection, and even of philosophy, to convince us, how little interest we should take in the greatest concerns of our neighbour, how little we should be affected by whatever relates to him, if the sense of propriety and justice did not correct the otherwise natural inequality of our sentiments.

Let us suppose that the great empire of China, with all its myriads of inhabitants, was suddenly swallowed up by an earthquake, and let us consider how a man of humanity in Europe, who had no sort of connexion with that part of the world, would be affected upon receiving intelligence of this dreadful calamity. He would, I imagine, first of all, express very strongly his sorrow for the misfortune of that unhappy people, he would make many melancholy reflections upon the precariousness of human life, and the vanity of all the labours of man, which could thus be annihilated in a moment. He would too, perhaps, if he was a man of speculation, enter into many reasonings concerning the effects which this disaster might produce upon the commerce of Europe, and the trade and business of the world in general. And when all this fine philosophy was over, when all these humane sentiments had been once fairly expressed, he would pursue his business or his pleasure, take his repose or his diversion, with the same ease and tranquillity, as if no such accident had happened. The most frivolous disaster which could befal himself would occasion a more real disturbance. If he was to lose his little finger to-morrow, he would not sleep to-night; but, provided he never saw them, he will snore with the most profound security over the ruin of a hundred millions of his brethren, and the destruction of that immense multitude seems plainly an object less interesting to him, than this paltry misfortune of his own. To prevent, therefore, this paltry misfortune to himself, would a man of humanity be willing to sacrifice the lives of a hundred millions of his brethren, provided he had never seen them? Human nature startles with horror at the thought, and the world, in its greatest depravity and corruption, never produced such a villain as could be capable of entertaining it. But what makes this difference? When our passive feelings are almost always so sordid and so selfish, how comes it that our active principles should often be so generous and so noble? When we are always so much more deeply affected by whatever concerns ourselves, than by whatever concerns other men; what is it which prompts the generous, upon all occasions, and the mean upon many, to sacrifice their own interests to the greater interests of others? It is not the soft power of humanity, it is not that feeble spark of benevolence which Nature has lighted up in the human heart, that is thus capable of counteracting the strongest impulses of self-love. It is a stronger power, a more forcible motive, which exerts itself upon such occasions. It is reason, principle, conscience, the inhabitant of the breast, the man within, the great judge and arbiter of our conduct. It is he who, whenever we are about to act so as to affect the happiness of others, calls to us, with a voice capable of astonishing the most presumptuous of our passions, that we are but one of the multitude, in no respect better than any other in it; and that when we prefer ourselves so shamefully and so blindly to others, we become the proper objects of resentment, abhorrence, and execration. It is from him only that we learn the real littleness of ourselves, and of whatever relates to ourselves, and the natural misrepresentations of self-love can be corrected only by the eye of this impartial spectator. It is he who shows us the propriety of generosity and the deformity of injustice; the propriety of resigning the greatest interests of our own, for the yet greater interests of others, and the deformity of doing the smallest injury to another, in order to obtain the greatest benefit to ourselves. It is not the love of our neighbour, it is not the love of mankind, which upon many occasions prompts us to the practice of those divine virtues. It is a stronger love, a more powerful affection, which generally takes place upon such occasions; the love of what is honourable and noble, of the grandeur, and dignity, and superiority of our own characters.

When the happiness or misery of others depends in any respect upon our conduct, we dare not, as self-love might suggest to us, prefer the interest of one to that of many. The man within immediately calls to us, that we value ourselves too much and other people too little, and that, by doing so, we render ourselves the proper object of the contempt and indignation of our brethren. Neither is this sentiment confined to men of extraordinary magnanimity and virtue. It is deeply impressed upon every tolerably good soldier, who feels that he would become the scorn of his companions, if he could be supposed capable of shrinking from danger, or of hesitating, either to expose or to throw away his life, when the good of the service required it.

One individual must never prefer himself so much even to any other individual, as to hurt or injure that other, in order to benefit himself, though the benefit to the one should be much greater than the hurt or injury to the other. The poor man must neither defraud nor steal from the rich, though the acquisition might be much more beneficial to the one than the loss could be hurtful to the other. The man within immediately calls to him, in this case too, that he is no better than his neighbour, and that by this unjust preference he renders himself the proper object of the contempt and indignation of mankind; as well as of the punishment which that contempt and indignation must naturally dispose them to inflict, for having thus violated one of those sacred rules, upon the tolerable observation of which depend the whole security and peace of human society. There is no commonly honest man who does not more dread the inward disgrace of such an action, the indelible stain which it would for ever stamp upon his own mind, than the greatest external calamity which, without any fault of his own, could possibly befall him; and who does not inwardly feel the truth of that great stoical maxim, that for one man to deprive another unjustly of any thing, or unjustly to promote his own advantage by the loss or disadvantage of another, is more contrary to nature, than death, than poverty, than pain, than all the misfortunes which can affect him, either in his body, or in his external circumstances.

When the happiness or misery of others, indeed, in no respect depends upon our conduct, when our interests are altogether separated and detached from theirs, so that there is neither connexion nor competition between them, we do not always think it so necessary to restrain, either our natural and, perhaps, improper anxiety about our own affairs, or our natural and, perhaps, equally improper indifference about those of other men. The most vulgar education teaches us to act, upon all important occasions, with some sort of impartiality between ourselves and others, and even the ordinary commerce of the world is capable of adjusting our active principles to some degree of propriety. But it is the most artificial and refined education only, it has been said, which can correct the inequalities of our passive feelings; and we must for this purpose, it has been pretended, have recourse to the severest, as well as to the profoundest philosophy.

Two different sets of philosophers have attempted to teach us this hardest of all the lessons of morality. One set have laboured to increase our sensibility to the interests of others; another, to diminish that to our own. The first would have us feel for others as we naturally feel for ourselves. The second would have us feel for ourselves as we naturally feel for others. Both, perhaps, have carried their doctrines a good deal beyond the just standard of nature and propriety.

The first are those whining and melancholy moralists, who are perpetually reproaching us with our happiness, while so many of our brethren are in misery, who regard as impious the natural joy of prosperity, which does not think of the many wretches that are at every instant labouring under all sorts of calamities, in the languor of poverty, in the agony of disease, in the horrors of death, under the insults and oppression of their enemies. Commiseration for those miseries which we never saw, which we never heard of, but which we may be assured are at all times infesting such numbers of our fellow-creatures, ought, they think, to damp the pleasures of the fortunate, and to render a certain melancholy dejection habitual to all men. But first of all, this extreme sympathy with misfortunes which we know nothing about, seems altogether absurd and unreasonable. Take the whole earth at an average, for one man who suffers pain or misery, you will find twenty in prosperity and joy, or at least in tolerable circumstances. No reason, surely, can be assigned why we should rather weep with the one than rejoice with the twenty. This artificial commiseration, besides, is not only absurd, but seems altogether unattainable; and those who affect this character have commonly nothing but a certain affected and sentimental sadness, which, without reaching the heart, serves only to render the countenance and conversation impertinently dismal and disagreeable. And last of all, this disposition of mind, though it could be attained, would be perfectly useless, and could serve no other purpose than to render miserable the person who possessed it. Whatever interest we take in the fortune of those with whom we have no acquaintance or connexion, and who are placed altogether out of the sphere of our activity, can produce only anxiety to ourselves, without any manner of advantage to them. To what purpose should we trouble ourselves about the world in the moon? All men, even those at the greatest distance, are no doubt entitled to our good wishes, and our good wishes we naturally give them. But if, notwithstanding, they should be unfortunate, to give ourselves any anxiety upon that account, seems to be no part of our duty. That we should be but little interested, therefore, in the fortune of those whom we can neither serve nor hurt, and who are in every respect so very remote from us, seems wisely ordered by Nature; and if it were possible to alter in this respect the original constitution of our frame, we could yet gain nothing by the change.

It is never objected to us that we have too little fellow-feeling with the joy of success. Wherever envy does not prevent it, the favour which we bear to prosperity is rather apt to be too great; and the same moralists who blame us for want of sufficient sympathy with the miserable, reproach us for the levity with which we are too apt to admire and almost to worship the fortunate, the powerful, and the rich.

Among the moralists who endeavour to correct the natural inequality of our passive feelings by diminishing our sensibility to what peculiarly concerns ourselves, we may count all the ancient sects of philosophers, but particularly the ancient Stoics. Man, according to the Stoics, ought to regard himself, not as something separated and detached, but as a citizen of the world, a member of the vast commonwealth of nature. To the interest of this great community, he ought at all times to be willing that his own little interest should be sacrificed. Whatever concerns himself, ought to affect him no more than whatever concerns any other equally important part of this immense system. We should view ourselves, not in the light in which our own selfish passions are apt to place us, but in the light in which any other citizen of the world would view us. What befalls ourselves we should regard as what befalls our neighbour, or, what comes to the same thing, as our neighbour regards what befalls us. 'When our neighbour,' says Epictetus, 'loses his wife, or his son, there is nobody who is not sensible that this is a human calamity, a natural event altogether according to the ordinary course of things; but, when the same thing happens to ourselves, then we cry out, as if we had suffered the most dreadful misfortune. We ought, however, to remember how we were affected when this accident happened to another, and such as we were in his case, such ought we to be in our own.' . . .

Source: https://www.marxists.org/reference/archive/smith-adam/works/moral/part03/part3b.htm. Accessed 12/6/2016.

PART II

Constitutional Foundations and Early U.S. History

For ... their mutual and general welfare

ARTICLES OF CONFEDERATION

1781

Introduction

After the Second Continental Congress issued a Declaration of Independence in 1776, it also set about drafting a document to govern relations among the thirteen states, which had declared their independence from Great Britain. It settled on a document, initially drafted by John Dickinson, known as the Articles of Confederation and proposed by Congress in 1777. The Articles did not formally go into effect until all the states ratified it, which did not occur until 1781.

As the title suggests, the Articles established what political scientists today call a confederal government, which divided power between a central and various constituent governments but in which the latter held primary power. Thus, Article II of the document proclaimed that "Each state retains its sovereignty, freedom, and independence, and every Power, Jurisdiction and right, which is not by this confederation expressly delegated to the United States, in Congress assembled." The Congress under the Articles of Confederation was unicameral, with each state delegation of from two to seven delegates having a single vote.

Nonetheless, the three passages below both refer, like later provisions in the U.S. Constitution, to the "general welfare," which, as reproduced in the U.S. Constitution, became a central "hook" that Congress would later use for disaster aid. Note that under Article IX of the Articles, it took the assent of 9 of 13 states in order for the Congress under the Articles to appropriate monies for "welfare" of the collective states.

Article III. The said states hereby severally enter into a firm league of friendship with each other, for their common defence, the security of their Liberties, and their mutual and general welfare, binding themselves to assist each other, against all force offered to, or attacks made upon them, or any of them, on account of religion, sovereignty, trade, or any other pretence whatever.

Article VIII. All charges of war, and all other expences that shall be incurred for the common defence or general welfare, and allowed by the united states in congress assembled, shall be defrayed out of a common treasury, which shall be supplied by the several states in proportion to the value of all land within each state, granted to or surveyed for any Person, as such land and the buildings and improvements thereon shall be estimated according to such mode as the united states in cong4ress assembled, shall from time to time direct and appoint. The taxes for paying that proportion shall be laid and levied by the authority and direction of the legislatures of the several states within the time agreed upon by the united states in congress assembled.

Article IX. The united states in congress assembled shall never engage in a war, nor grant letters of marque and reprisal in time of peace, nor enter into any treaties or alliances, nor coin money, nor regulate the value thereof, nor ascertain the sums and expences necessary for the defence and welfare of the united states, or any of them, nor emit bills, nor borrow money on the credit of the united states, nor appropriate money, nor agree upon the number of vessels of war, to be built or purchased, or the number of land or sea forces to be raised, nor appoint a commander in chief of the army or navy, unless nine states assent to the same....

"[A] specification of the objects alluded to by these general terms immediately follows"

James Madison as Publius
THE FEDERALIST No. 41

1788

Introduction

The Federalist, which was eventually published as a two-volume book, began as a series of essays in New York newspapers designed to advocate adoption of the new Constitution. Alexander Hamilton, James Madison, and John Jay authored the essays under the pen name of Publius. This essay, written by James Madison, dealt in part with charges that the new Constitution, and especially the taxing and spending clause, vested Congress with too much power. Throughout his career, Madison interpreted this clause restrictively so that Congress could only tax and spend for projects enumerated or implied by other listed powers. This interpretation provided the basis for his veto of the Bonus Bill of 1817, which Congress had designated for internal improvements like roads and canals.

[*The initial portion of this essay has been omitted.*]

Some, who have not denied the necessity of the power of taxation, have grounded a very fierce attack against the Constitution, on the language in which it is defined. It has been urged and echoed, that the power "to lay and collect taxes, duties, imposts, and excises, to pay the debts, and provide for the common defense and general welfare of the United States," amounts to an unlimited commission to exercise every power which may be alleged to be necessary for the common defense or general welfare. No stronger proof could be given of the distress under which these writers labor for objections, than their stooping to such a misconstruction. Had no other enumeration or definition of the powers of the Congress been found in the Constitution, than the general expressions just cited, the authors of the objection might have had some color for it; though it would have been difficult to find a reason for so awkward a form of describing an authority to legislate in all possible cases. A power to destroy the freedom of the press, the trial by jury, or even to regulate the course of descents, or the forms of conveyances, must be very singularly expressed by the terms "to raise money for the general welfare."

But what color can the objection have, when a specification of the objects alluded to by these general terms immediately follows, and is not even separated by a longer pause than a semicolon? If the different parts of the same instrument ought to be so expounded, as to give meaning to every part which will bear it, shall one part of the same sentence be excluded altogether from a share in the meaning; and shall the more doubtful and indefinite terms be retained in their full extent, and the clear and precise expressions be denied any signification whatsoever? For what purpose could the enumeration of particular powers be inserted, if these and all others were meant to be included in the preceding general power? Nothing is more natural nor common than first to use a general phrase, and then to explain and qualify it by a recital of particulars.

But the idea of an enumeration of particulars which neither explain nor qualify the general meaning, and can have no other effect than to confound and mislead, is an absurdity, which, as we are reduced to the dilemma of charging either on the authors of the objection or on the authors of the Constitution, we must take the liberty of supposing, had not its origin with the latter. The objection here is the more extraordinary, as it appears that the language used by the convention is a copy from the articles of Confederation. The objects of the Union among the States, as described in article third, are "their common defense, security of their liberties, and mutual and general welfare. " The terms of article eighth are still more identical: "All charges of war and all other expenses that shall be incurred for the common defense or general welfare, and allowed by the United States in Congress, shall be defrayed out of a common treasury," etc. A similar language again occurs in article ninth. Construe either of these articles by the rules which would justify the construction put on the new Constitution, and they vest in the existing Congress a power to legislate in all cases whatsoever.

But what would have been thought of that assembly, if, attaching themselves to these general expressions, and disregarding the specifications which ascertain and limit their import, they had exercised an unlimited power of providing for the common defense and general welfare? I appeal to the objectors themselves, whether they would in that case have employed the same reasoning in justification of Congress as they now make use of against the convention. How difficult it is for error to escape its own condemnation!

PUBLIUS.

Source: Avalon Project at Yale University.

To promote the general welfare

U.S. CONSTITUTION AND BILL OF RIGHTS

1789

Introduction

A Convention of 55 men from 12 of the 13 states met at the State House (today's Independence Hall) in Philadelphia during the summer of 1787. Although many delegates arrived expecting to reform the existing Articles of Confederation, Virginia's James Madison was among the delegates who thought that it would be better to start with a clean slate. Early in the proceedings, Governor Edmund Randolph of Virginia introduced the Virginia Plan, which proposed replacing the Articles with a federal system of government which divided the central government into three branches, in which Congress would be bicameral, and in which states would be represented according to population. After numerous compromises between more and less populous states, free states and slave states, and those who wanted alterations in one or more branches of the proposed government or in the relation between the national government and the states, delegates finally settled on a document of seven articles. State conventions ratified the document, which went into effect in 1789, and which 27 amendments have subsequently modified.

The opening paragraph of the document set forth the purposes of the document, which included promoting "the general Welfare." Article I, Section 8 further outlined the powers of Congress, which was vested with what Alexander Hamilton called "the power of the purse." This included the power to tax and spend on behalf of the general welfare.

In contrast to the prior Articles of Confederation, the new Constitution vested Congress with powers over interstate, foreign, and Indian commerce. This clause could justify attempts to facilitate commerce by mitigating possible obstructions that might be caused or heightened by natural disasters.

The line between natural catastrophes and catastrophes caused by war and rebellion is often a fine one. Significantly, Article I, Section 8 also vested Congress with the power to call out the Militia to "suppress Insurrections and repel Invasions." Both state and national governments currently use the National Guard to deal with natural disasters. In a similar fashion, Article I, Section 9 provides for suspension of the Writ of Habeas Corpus in times of rebellion or invasion.

The last clause of Article I, Section 8, contains the necessary and proper clause, which serves as the basis of the doctrine of implied powers. This clause can be used in cases where disaster aid is considered essential to national defense, to the restoration of interstate commerce, or to other powers enumerated in the Constitution. Although it is phrased in terms of a limitation, a provision in Article I, Section 10 furthers federalism by enabling states to enter into interstate compacts, some of which are now used in emergency situations, with congressional consent.

The president often calls upon the military in times of natural disasters. His control is associated with civilian control and is a guard against the kind of military coups that have occurred in many others presidential systems. Today's militia, the National Guard, is under state command, unless federalized by the president for a specific emergency. Although it is directed to preserving "republican" government, Article IV, Section 4 of the Document arguably provides something of a model for state and federal cooperation by establishing that it will take an application by the state executive or legislature to trigger a national response.

This clause should be considered in conjunction with the Tenth Amendment, which was proposed by the first Congress in 1789 and ratified by the states in 1791 as part of what is today known as the Bill of Rights. This amendment serves as the basis for the doctrine of state police powers by reserving powers not delegated to the national government to the states. The takings clause of the Fifth Amendment is also included here because it affects the governmental regulation and property often even for purposes of prevention and mitigation.

Preamble

We the People of the United States, in Order to form a more perfect Union, establish Justice, insure domestic Tranquility, provide for the common defence, promote the general Welfare, and secure the Blessings of Liberty to ourselves and our Posterity, do ordain and establish this Constitution for the United States of America.

Article I, Section 8, para. 1

The Congress shall have Power to lay and collect Taxes, Duties, Imposts and Excises, to pay the Debts and provide for the common Defense and general Welfare of the United States.

Article I, Section 8, para. 3

[The Congress shall have Power] To regulate Commerce with foreign Nations, and among the several states, and with the Indian Tribes.

Article I, Section 8, para. 15-16

[The Congress shall have Power] To provide for calling forth the Militia to execute the Laws of the Union, suppress Insurrections and repel invasions; To Provide for organizing, arming and disciplining, the Militia, and for governing such part of them as may be employed in the Service of the United States, reserving to the States respectively, the appointment of the Officers, and the Authority of training the Militia according to the discipline prescribed by Congress.

Article I, Section 8, para. 17

[The Congress shall have Power] to make all Laws which shall be necessary and proper for carrying into Execution the foregoing Powers, and all other Powers vested by this Constitution in the Government of the United States, or in any Department or Officer thereof.

Article I, Section 9, Cl. 2

The Privilege of the Writ of Habeas Corpus shall not be suspended, unless when in Cases of Rebellion or Invasion the public Safety may require it.

Article I, Section 10

No State shall, without the Consent of Congress ... enter into any Agreement or Compact with another State ....

Article II, Section 2

The President shall be Commander in Chief of the Army and Navy of the United States, and of the Militia of the several States, when called into the actual Service of the United States.

Article IV, Section 4

The United States shall guarantee to every State in this Union a Republican Form of Government, and shall protect each of them against Invasion; and on Application of the Legislature, or of the Executive (when the Legislature cannot be convened) against domestic Violence.

The Bill of Rights

Amendment V

... nor shall private property be taken for public use, without just compensation.

Amendment X

The powers not delegated to the United States by the Constitution, nor prohibited by it to the States, are reserved to the States respectively, or to the people.

Misfortunes every part was liable to,
and our own turn might be next

King George III in Parliament
AID TO THE CARIBBEAN

1790

Introduction

Although it is pretty much taken for granted that modern governments will provide aid to their citizens in times of natural disasters, it was not necessarily always so. This appears to be one of the earliest cases (Britain had also responded with aid after a fire in Charleston, South Carolina in 1740) where the British government responded to a hurricane disaster in the Caribbean by providing such aid. Notable is the attempt to assess where aid was most needed and the attempt to provide aid without undermining self-help or private charity. Politically, this aid was a good move in light of the fact that the Caribbean colonies had not joined the thirteen colonies to the North in seeking separation from the mother country.

On Tuesday Jan. 23d the House met for the first time since their adjournment, when the petitions aforesaid were presented, and Lord North signified his majesty's approbation to the same; and the next day the House having resolved itself into a committee of supply, his lordship rose and expatiated on the nature of the misfortunes these islands had sustained by the late hurricanes, which he said had reduced them to the lowest state of wretchedness, and well intitled them to some assistance from the parent state; but it was not expected that the benevolence of this country should extend to a total indemnification for their losses; nor were the propositions he should offer formed on that idea. This calamity was a private hardship, it was true, pressing more immediately on one part of the empire; but misfortunes every part was liable to, and our own turn might be next; it could not therefore be thought reasonable that the people of this country should entirely exonerate the West Indians from the burthen of their distress. He intended to propose nothing more than a relief to the poorer class of sufferers, who were least able to support or retrieve their losses, a gift merely of benevolence, and not of reparation. As to the more opulent inhabitants, though we might, and should sympathize with their distress, circumstances considered, it would not be prudent to think of offering them any adequate relief. On this view of the business, it would be possible to do all that the house intended, without any evidence to ascertain the exact amount of the damage done: for that some instant relief was requisite, public notoriety sufficiently proved, and they might exercise their judgment in determining the quantum.

His lordship next took a comparative view of the losses sustained in the two islands of Jamaica and Barbados. In the former, two parishes only were wasted, but the whole island of Barbados was destroyed. Were the amount of the loss in point of property destroyed, however, only to be considered, or were the gift to be in proportion to the value of their commerce with this country, the two parishes in Jamaica which were injured, might claim more regard than the whole island of Barbados; but on the principle of benevolence the latter should undoubtedly have the largest supply; for in Jamaica very few but opulent planters had suffered, and in the other island there were vast numbers of poor entirely ruined. Besides this, it was to be remembered that as a small part only of Jamaica had lost much by the hurricane, the remainder of that country could afford much relief to their indigent neighbours, and be able at least to supply them with viands; while in Barbados the whole crop was destroyed, and the entire face of the country made desolate, so that the miserable poor could expect no aid from internal charity.

These reasons, his lordship said, had gone so far with some gentlemen, that they would have had the gift to one island four times as large as that to the other; but thinking that proportion rather unreasonable, he should only propose that Barbados should have twice as much as Jamaica. With respect to the sum proper to be given to each, and the mode of distribution, he was of opinion, that part should be sent over in money, part in provisions and lumber, in such proportions as should be deemed best adapted to the exigencies of the sufferers, which when arrived at the islands, should be distributed there in the most impartial manner by the vestries, assisted by the governors and council; for the particular necessities of each inhabitant could only be known on the spot.l

His lordship concluded by moving, that 80,000l, be granted to his majestry, to be disposed of as to his majesty shall seem best, for the relief of the island of Barbados; and for that of Jamaica, 40,000l.

After some conversation between his lordship, Mr. Estwick, Mr. Pulteney, and Mr. Burke, in which the latter suggested than an open committee of West-India merchants in London might best direct the Treasury in the application of the money, these motions were agreed to unanimously.

Sir William Guise hinted that it would be proper these gifts should be issued from the treasury without fee or deduction; with this lord North immediately concurred, and, framing another esolution to that effect, it was also agreed to.

'Tis unnessary to add that the Upper-House will, undoubtedly, pass this bill into an act with equal dispatch and humanity: nothing therefore now remains, but for the public to imitagte the exemplary conduct of parliament, and to promote, in conjunction with the following named committee, that plan of national benevolence, regard, and protection, which must do honour to themselves, and be rewarded with the blessings of thousands.

[A long list of names follows]

Source: John Fowler, *A General Account of the Calamities Occasioned by the Late Tremendous Hurricanes and Earthquakes in the West India Islands, Foreign as Well as Domestic...* (London: Printed for J. Stockdale, 1781), pp. 80-83.

Existing or threatening calamities, the just judgments of God against prevalent iniquity

---

John Adams
PROCLAMATION OF DAY OF FASTING, HUMILIATION AND PRAYER

(March 23, 1798)

Introduction

Although it is easy to idealize the early American republic, the nation faced serious problems. These included the degree to which it was bound by an alliance with the French, who had helped the U.S. win the Revolutionary War, in their ongoing conflict with Britain. This conflict was exacerbated by the ideology of the French Revolution, which had called for "liberty, equality, and fraternity" and which was strongly anti-clerical.

Both Britain and France sought to strengthen their own position by forbidding trade with their rival, and American ships were often the victims of such orders. Although engaged in an undeclared war at sea, President Adams did not call, and congress did not declare, a formal war. In this proclamation, Adams sought to follow colonial precedents, which were especially strong in his native New England, of calling for a national day of fasting, humiliation, and prayer (President Washington had issued two prior Thanksgiving Proclamations).

Some Democratic-Republicans questioned whether such proclamations violated the disestablishment clause of the First Amendment. Others undoubtedly worried that this was Adams's way of identifying the French and Democratic-Republicans with infidelity, a charge that Federalists would repeat in the critical election of 1800.

Although this proclamation aimed at avoiding the evils that would flow from a war between France and the United

States, it follows earlier proclamations in associating such disastrous events with God's punishment of the people's sins.

As the safety and prosperity of nations ultimately and essentially depend on the protection and the blessing of Almighty God, and the national acknowledgment of this truth is not only an indispensable duty which the people owe to Him, but a duty whose natural influence is favorable to the promotion of that morality and piety without which social happiness can not exist nor the blessings of a free government be enjoyed; and as this duty, at all times incumbent, is so especially in seasons of difficulty or of danger, when existing or threatening calamities, the just judgments of God against prevalent iniquity, are a loud call to repentance and reformation; and as the United States of America are at present placed in a hazardous and afflictive situation by the unfriendly disposition, conduct, and demands of a foreign power, evinced by repeated refusals to receive our messengers of reconciliation and peace, by depredation on our commerce, and the infliction of injuries on very many of our fellow-citizens while engaged in their lawful business on the seas—under these considerations it has appeared to me that the duty of imploring the mercy and benediction of Heaven on our country demands at this time a special attention from its inhabitants.

I have therefore thought fit to recommend, and I do hereby recommend, that Wednesday, the 9th day of May next, be observed throughout the United States as a day of solemn humiliation, fasting, and prayer; that the citizens of these States, abstaining on that day from their customary worldly occupations, offer their devout addresses to the Father of Mercies agreeably to those forms or methods which they have severally adopted as the most suitable and becoming; that all religious congregations do, with the deepest humility, acknowledge before God the manifold sins and transgressions with which we are justly chargeable as individuals and as a nation, beseeching Him at the same time, of His infinite grace, through the Redeemer of the World, freely to remit all our offenses, and to incline us by His Holy Spirit to that sincere repentance and reformation which may afford us reason to hope for his inestimable favor and heavenly benediction; that it be made the subject of particular and earnest supplication that our country may be protected from all the dangers which threaten it; that our civil and religious privileges may be preserved inviolate and perpetuated to the latest generations; that our public councils and magistrates may be especially enlightened and directed at this critical period; that the American people may be united in those bonds of amity and mutual confidence and inspired with that vigor and fortitude by which they have in times past been so highly distinguished and by which they have obtained such invaluable advantages; that the health of the inhabitants of our land may be preserved, and their agriculture, commerce, fisheries, arts, and manufactures be blessed and prospered; that the principles of genuine piety and sound morality may influence the minds and govern the lives of every description of our citizens, and that the blessings of peace, freedom, and pure religion may be speedily extended to all the nations of the earth.

And finally, I recommend that on the said day the duties of humiliation and prayer be accompanied by fervent thanksgiving to the Bestower of Every Good Gift, not only for His having hitherto protected and preserved the people of these United States in the independent enjoyment of their religious and civil freedom, but also for having prospered them in a wonderful progress of population, and for conferring on them many and great favors conducive to the happiness and prosperity of a nation.

Given under my hand and the seal of the United States of America, at Philadelphia, this 23d day of March, A. D. 1798, and of the Independence of the said States the twenty-second.

JOHN ADAMS.
By the President:
TIMOTHY PICKERING,
Secretary of State.

Source: Website of the Miller Center, the University of Virginia.

Population, when unchecked, increases in a geometrical ratio. Subsistence increases only in an arithmetical ratio

---

Thomas Malthus
AN ESSAY ON THE PRINCIPLE OF POPULATION

1798

Introduction

Of all the eighteenth and nineteenth century economists, none had a darker more apocalyptic vision than Thomas Robert Malthus (1766-1834), an English clergyman who saw an inevitable tie between population growth and human misery. If such misery is a necessary concomitant of population growth, one could conclude that natural disasters, especially diseases and famines, are natural means of attaining stability, and that human interference (as with the poor laws that he criticized) will only postpone the day of reckoning. Social Darwinists later advocated this view.

From Chapter I

I think I may fairly make two postulate.

First, That food is necessary to the existence of man.

Secondly, That the Passion between the sexes is necessary, and will remain nearly in its present state. (p. 11)

Affirming then, my postulate as granted, I say that the power of population is indefinitely greater than the power in the earth to produce subsistence for man.

Population, when unchecked, increases in a geometrical ratio. Subsistence increases only in an arithmetical ratio. A slight acquaintance with numbers will shew the immensity of the first power in comparison of the second.

By that law of our nature which makes food necessary to the life of man, the effects of these two unequal powers must be kept equal.

This implies a strong and constantly operating check on population from the difficulty of subsistence. This difficulty must fall some where, and must necessarily be severely felt by a large portion of mankind. (13-14)

This natural inequality of the two powers of population, and of production in the earth, and that great law of our nature which must constantly keep their effects equal, form the great difficulty that to me appears insurmountable in the way to the perfectibility of society. All other arguments are of slight and subordinate consideration in comparison of this. I see no way by which man can escape from the weight of this law which pervades all animated nature. No fancied equality, no agrarian regulations in their utmost extent, could remove the pressure of it even for a single century. And it appears, therefore, to be decisive against the possible existence of a society, all the members of which, should live in safe, happiness, and comparative leisure; and fell no anxiety about providing the means of subsistence for themselves and families.

Consequently, if the premises are just, the argument is conclusive against the perfectibility of the mass of mankind. (16-17)

From Chapter 5

The poor-laws of England were undoubtedly instituted for the most benevolent purpose; but there is great reason to think that they have not succeeded in their intention. They certainly mitigate some cases of very severe distress which might otherwise occur; yet the state of the poor who are supported by parishes, considered in all its circumstances, is very far from being free from misery. But one of the principal objections to them is, that for this assistance which some of the poor receiv e, in itself almost a doubtful blessing, the whole class of the common people of England, is subjected to a set of grating, inconvenient, and tyrannical laws, totally inconsistent with the genuine spirit of the constitution. (91-92)

From Chapter 7

Mr. Pitt's Poor-bill has the appearance of being framed with benevolent intentions, and the clamour raised against it was in many respects ill directed, and unreasonable. But it must be confessed that it possesses in a high degree the great and radical defect of all systems of the kind, that, of tending to increase population without increasing the means for its support, and thus to depress the condition of those that are not supported by parishes, and consequently, to create moor poor (140-141).

Famine seems to be the last, the most dreadful resource of nature. The power of population is so superior to the power in the earth to produce subsistence for man, that premature death must in some shape or other visit the human race. The vices of mankind are active and able ministers of depopulation. They are the precursors in the great army of destruction; and often finish the dreadful work themselves. But should they fail in this war of extermination, sickly, seasons, epidemics, pestilence, and plague, advance in terrific array, and sweep off their thousands and ten thousands. Should success be still incomplete; gigantic inevitable family stalks in the rear, and with one might blow, levels the population with the food of the world.

Must it not then be acknowledged by an attentive examiner of the histories of mankind, that in every age and every State in which man has existed, or does now exist,

That the increase of population is necessarily limited by the means of subsistence.

That population does invariably increase when the means of subsistence increase. And,

That the superior power of population is repressed, and the actual population kept equal to the means of subsistence by misery and vice. (139-141)

Source: Thomas Malthus, *An Essay on the Principle of Population, as it affects the Future Improvement of Society, with remarks on the Speculations of Mr. Godwin, M. Condorcet, and Other Writers.* London: J. Johnson, 1798. Note, I have modernized spellings, to replace f's, where appropriate, with s's.

[S]uspended ... the collection of bonds due ...
by merchants ... who have suffered

BILL FOR RELIEF OF PORTSMOUTH

1803

Introduction

In 1803, the city of Portsmouth, New Hampshire, suffered from a severe fire that destroyed 114 buildings. About $45,000 in charitable gifts were collected. In addition, Congress adopted the following resolution, which provided the suspension of bond payments from merchants who had been adversely affected by the fire. Although an indirect, rather than a direct appropriation, this relief is significant in part because it was authorized when both elected branches of the national government were dominated by Democratic-Republicans, who were generally wary of national aid to the states for local issues. Hrafnkell Haroldsson claims that this is the first example of federal disaster relief.

Be it enacted, by the Senate and House of Representatives of the United States of America, in Congress assembled, That the secretary of the treasury be, and he hereby is authorized and directed to cause to be suspended for months, the collection of bonds due to the United States by merchants of Portsmouth, in New Hampshire, who have suffered by the late conflagration of that town.

Source: Document found at *A Century of Lawmaking for a New Nation: U.S. Congressional Documents and Debates, 1776-1875*. Bill 11 of 54. https://memory.loc.gov/cgi-bin/ampage.

Source: Hrafnkell Haraldsson, "Phil Robertson Dishonestly Claims Founding Fathers Opposed Federal Disaster Relief," September 9, 2016. Found at http://www.politicususa.com/2016/09/09/phil-robertson-dishonestly-claims-founding-fathe... Accessed 12/5/2016.

[O]rder the insurgents to disperse and retire peaceably to their abodes within a limited time

---

THE INSURRECTION ACT

1817

Introduction

Acts of lawlessness can create situations equivalent to natural disasters. The Insurrection Act, originally enacted in 1817, and subsequently amended, grants the president the right to respond to insurrections with military force. The law, as amended, seeks to balance the federal responsibility to respond to requests for aid, with state primacy in the area of law enforcement and civilian control over the military.

Today's law is found in 10 U.S. Code, Sections 331-335. These sections should be understood within the context of the Posse Comitatus Act of 1878, which limits the use of the military for routine law enforcement functions that are typically carried out at the state level.

The President has used this act to quell the Los Angeles riots in 1992 and to suppress looting in St. Croix, Virgin Islands after Hurricane Hugo in 1989 (Elsea, 2005, p. 3).

Section 331

Whenever there is an insurrection in any State against its government, the President may, upon the request of its legislature or of its governor if the legislature cannot be convened, call into Federal service such of the militia of the other States, in the number requested by that State, and use such of the armed forces, as he considers necessary to suppress the insurrection.

Section 332

Whenever the President considers that unlawful obstructions, combinations, or assemblages, or rebellion against the authority of the United States, make it impracticable to enforce the laws of the United States in any State by the ordinary course of judicial proceedings, he may call into Federal service such of the militia of any State, and use such of the armed forces, as he consider necessary to enforce those laws or to suppress the rebellion.

Section 333

The president, by using the militia or the armed forces, or both, or by any other means shall take such measures as he considers necessary to suppress, in a State, any insurrection, domestic violence, unlawful combination, or conspiracy, if it—

(1) so hinders the execution of the laws of that State, and of the United States within the State, that any part or class of its people is deprived of a right, privilege, immunity, or protection named in the Constitution and secured by law, and the constituted authorities of that State are unable, fail, or refuse to protect that right, privilege, or immunity, or to give that protection; or

(2) opposes or obstructs the execution of the laws of the United States or impedes the course of justice under those laws.

In any situation covered by clause (1), the State shall be considered to have denied the equal protection of the laws secured by the Constitution.

Section 334

Whenever the President considers it necessary to use the militia or the armed forces under this chapter, he shall, by proclamation, immediately order the insurgents to disperse and retire peaceably to their abodes within a limited time.

Section 335

For purposes of this chapter, the term "State" includes Guam and the Virgin Islands.

Source: U.S. Code at https://www.law.cornell.edu/uscode/texgt/10/331.

Nor is there any danger that such a power will be abused while the vigor of representative responsibility remains unimpaired

---

St. George Tucker
HOUSE REPORT ON ROADS AND CANALS

December 15, 1817

Introduction

After James Madison vetoed a bill providing for funding for roads, the House of Representatives formed a committee, headed by St. George Tucker of Virginia, which issued a report in which it cited numerous precedents for such construction and for broad congressional powers under the taxing and spending clause. The section excerpted below ends with a summary of the former and focuses on the latter arguments. Note how Tucker's reports looks back to the breadth of past congressional expenditures, including aid to individuals in need, for building a case for other such expenditures.

It is true that the wants of the Union cannot confer power under the constitution, but they may justly be touched upon as affording aid in its construction. They may have been clearly foreseen, and must have been supposed to be provided for. If the power to carry on war implies "the necessary and proper" means of conducting it to a safe and prosperous issue, and if, without the use of these means, the burdens, and the privations, and the miseries of war are to be indefinitely increased, and its issue (always doubtful) rendered yet more precarious and unprosperous, are we not justified in presuming these means to have been contemplated as being vested in the General Government? Are we not justified in asserting this "necessary" power – the power of constructing roads and canals, at least *with the assent of the States*.

If your committee have not erred in attributing to Congress a constitutional power to make roads and canals, either as an original or accessory power, it would seem that no doubt could remain of the right of applying our revenues to those purposes. If, indeed, the power was denied to the General Government of constructing roads and canals themselves, a question might still arise whether it had not power to appropriate part of the revenue "to aid in the construction of roads and canals by the States?"

There is, perhaps, no part of the constitution more unlimited than that which relates to the application of the revenues which are to be raised under its authority. The power is given "to lay and collect taxes, &c; to pay the debts, and provide for the common defence and general welfare of the United States;" and though it be readily admitted that, as this clause is only intended to designate the objects for which revenue is to be raised, it cannot be construed to extend the specified powers of Congress, yet it would be difficult to reconcile either the generality of the expression, or the course of administration under it, with the idea that Congress has not a discretionary power over its expenditures, limited only by their application "to the common defense and general welfare."

A few of the very great variety of instances in which the revenues of the United States have been applied to objects not falling within the specified powers of Congress, or those which may be regarded as incidental to them, will best illustrate this remark.

Thus, it can scarcely be conceived that, if construed with rigor, the constitution has conferred the power to purchase a library, either specifically or as a "necessary" incident to legislation; still less, perhaps can the pious services of a chaplain, or the purchase of expensive paintings for ornamenting the hall of session, or various other expenditures of a similar character, be considered as "necessary" incidents to the power of making laws. Yet to these and to similar objects have the funds of the United States been freely applied at every successive session of congress, without a question as to the constitutionality of the application.

It would be more difficult to reduce, under the specific or accessory powers of Congress, the liberal donations to the wretched sufferers in Venezuela, or the employment of our revenues in the useful and interesting enterprise to the Pacific.

The bounties allowed for the encouragement of the fisheries form another expenditure that does not fall under any of the powers granted by the constitution, nor could it fairly be considered as inferrible from the powers granted upon the strict principles sometimes contended for. The same objections would apply to actual bounties paid to manufacturers for their encouragement, and to the indirect encouragement given to them, and which operates as a bounty to one class of the community, and as a *tax* upon the rest. These, and a variety of other appropriations, can only be justified upon the principle that the general clause in question has vested in Congress a discretionary power to use for the "general welfare" the funds which they are authorized to raise.

Nor is there any danger that such a power will be abused while the vigor of representative responsibility remains unimpaired. It is on this principle that the framers of the constitution mainly relied for the protection of the public purse. It was a safe reliance. It was manifest that there was no other subject on which representative responsibility would be so great. On the other hand, while this principle was calculated to prevent abuses in the appropriations of public money, it was equally necessary to give an extensive discretion to the legislative body in the disposition of the revenues; since no human foresight could discern, nor human industry enumerate, the infinite variety of purposes to which the public money might advantageously and legitimately be applied. The attempt would have been to *legislate*, not to frame a constitution; to foresee and provide specifically for the wants of future generations, not to frame a rule of conduct for the legislative body. Hence proceeds the use of this general phrase in relation to the purposes to which the revenues may be applied, whilst the framers of the instrument, in the clause which concludes the *enumeration of powers*, scrupulously avoid the use of so comprehensive an expression, and confine themselves to the grant of such incidental power as might be both "necessary and proper" to the exercise of the specified powers.

Nor is it conceived that this construction of the constitution is calculated to give that unlimited extent to the powers of the Federal Government which by some seems to have been apprehended. There is a distinction between the power to appropriate money for a purpose, and the power to do the act for which it is appropriated; and, if so, the authority to appropriate money "for the general welfare" does not, by fair construction, extend the specified or incidental powers of Government. Thus, in the case under consideration, if the power to make a road or dig a canal is not given, the power of appropriating money cannot confer it, however generally it may be expressed. If there were no other limitation, the rights of the respective States over their soil and territory would operate as a restriction.

Whilst this appears to be safe as well as fair construction of the constitution, it is also that which has been practically given to it since the origin of the Government. Of this, the instances already mentioned furnish some evidence; and it is apprehended that, upon the rigid principles of construction asserted both in regard to the enumeration of powers and the appropriation of revenue, the acts of the Federal Government, including all its branches, will exhibit a continued series of violations of the constitution from the first session after its adoption to the present day.

It would behoove us to turn over the statute book, and deliberately examine how, upon these principles, the laws giving bounties to fishermen, encouraging manufacturers, establishing trading-houses with the Indians, erecting and constructing beacons, piers, and light-houses, purchasing libraries, adorning with paintings the chambers of Congress, giving charity to suffering foreigners, constructing roads through the different States, and establishing banks, can be reconciled to the provisions of the constitution. If, as has been remarked by high authority, the constitutional question can be "precluded by repeated recognitions, under varied circumstances, of the validity 'of the exercise of a power by Congress' in acts of the legislative, executive, and judicial branches of the Government, accompanied by indications, in different modes, of a concurrence of a general will of the nation," the advocates for those powers in the General Government can find little difficult in supporting the pretention.

From all these considerations, your committee submit it as their opinion that congress has the constitutional power to construct roads and canals through the several States, *with the assent of the States,* on such terms as may be agreed to, leaving the *jurisdictional rights* in the States, respectively....

Source: *American State Papers, Miscellaneous,* Vol. 2, pp. 446-447. Https://memory.loc.gov/cgi-bin/ampage?collId=llsp&fileName=038/llsp03.db&Page=443.

I am of opinion that Congress do not possess this power

President James Monroe
VETO OF THE CUMBERLAND ROAD BILL

May 4, 1822

Introduction

Democratic Republicans like Thomas Jefferson and James Madison generally argued for what they called "strict construction" of the U.S. Constitution. They were concerned that more liberal construction would destroy the doctrine that tied congressional exercises of power to those that the Constitution enumerated. James Monroe, who had originally opposed the U.S. Constitution because of similar concerns over the breadth of power that it vested in the national government, expressed similar concerns in this veto address of the Cumberland Road Bill. Like Jefferson and Madison before him, both of whom had opposed the constitutionality of the national bank, Monroe thought that expenditures for internal improvements benefited some areas of the country over others and thus did not qualify as proper expenditures for the "general welfare." This issue was quite similar to concerns expressed over expenditures on disaster relief. Implicit in such arguments (and similar to current concerns over rising federal budget deficits) is fear that the Constitution provided no logical stopping point.

To the House of Representatives:

Having duly considered the bill entitled "An act for the preservation and repair of the Cumberland road," it is with deep regret, approving as I do the policy, that I am compelled to object to its passage and to return the bill to the House of Representatives, in which it originated, under a conviction that Congress do not possess the power under the Constitution to pass such a law.

A power to establish turnpikes with gates and tolls, and to enforce the collection of tolls by penalties, implies a power to adopt and execute a complete system of internal improvement. A right to impose duties to be paid by all persons passing a certain road, and on horses and carriages, as is done by this bill, involves the right to take the land from the proprietor on a valuation and to pass laws for the protection of the road from injuries, and if it exist as to one road it exists as to any other, and to as many roads as Congress may think proper to establish. A right to legislate for one of these purposes is a right to legislate for the others. It is a complete right of jurisdiction and sovereignty for all the purposes of internal improvement, and not merely the right of applying money under the power vested in Congress to make appropriations, under which power, with the consent of the States through which this road passes, the work was originally commenced, and has been so far executed. I am of opinion that Congress do not possess this power; that the States individually can not grant it, for although they may assent to the appropriation of money within their limits for such purposes, they can grant no power of jurisdiction or sovereignty by special compacts with the United States. This power can be granted only by an amendment to the Constitution and in the mode prescribed by it.

If the power exist, it must be either because it has been specifically granted to the United States or that it is incidental to some power which has been specifically granted. If we examine the specific grants of power we do not find it among them, nor is it incidental to any power which has been specifically granted.

It has never been contended that the power was specifically granted. It is claimed only as being incidental to some one or more of the powers which are specifically granted. The following are the powers from' which it is said to be derived:

First, from the right to establish post-offices and post-roads; second, from the right to declare war; third, to regulate commerce; fourth, to pay the debts and provide for the common defense and general welfare; fifth, from the power to make all laws necessary and proper for carrying into execution all the powers vested by the Constitution in the Government of the United States or in any department or officer thereof; sixth and lastly, from the power to dispose of and make all needful rules and regulations respecting the territory and other property of the United States.

According to my judgment it can not be derived from either of those powers, nor from all of them united, and in consequence it does not exist.

Having stated my objections to the bill, I should now cheerfully communicate at large the reasons on which they are founded if I had time to reduce them to such form as to include them in this paper. The advanced stage of the session renders that impossible. Having at the commencement of my service in this high trust considered it a duty to express the opinion that the United States do not possess the power in question, and to suggest for the consideration of Congress the propriety of recommending to the States an amendment to the Constitution to vest the power in the United States, my attention has been often drawn to the subject since, in consequence whereof I have occasionally committed my sentiments to paper respecting it. The form which this exposition has assumed is not such as I should have given it had it been intended for Congress, nor is it concluded. Nevertheless, as it contains my views on this subject, being one which I deem of very high importance, and which in many of its bearings has now become peculiarly urgent, I will communicate it to Congress, if in my power, in the course of the day, or certainly on Monday next.

JAMES MONROE.

Source: James Monroe, "Veto Message," May 4, 1822. Online by Gerhard Peters and John T. Woolley, *The American Presidency Project.* http://www.presidency.ucsb.edu/ws/?pid=66322.

A power to lay taxes for any purposes whatsoever is a general power; a power to lay taxes for certain specified purposes is a limited power

---

Joseph Story
COMMENTARIES ON THE CONSTITUTION
Vol. 2
1833

Introduction

Few, if any, constitutional commentators have done more to explain and interpret the Constitution than Joseph Story (1779-1845), who served as an associate justice of the U.S. Supreme Court from 1811-1845. A Harvard professor, Story was known for explaining and popularizing the U.S. Constitution not only to law students but also to the general public.

This interpretation of the general welfare clause was one of the most influential of its day and continues to guide modern understandings of the provision. Story, who began political life as a Federalist but was appointed to the Supreme Court by James Madison, who was a Democratic-Republican, charted a course between those like Alexander Hamilton, who believed the taxing and spending clause was an unlimited power, and those, like James Madison and Thomas Jefferson, who thought that it offered Congress no additional authority beyond other enumerated powers.

§ 904. Before proceeding to consider the nature and extent of the power conferred by this clause, and the reasons, on which it is founded, it seems necessary to settle the grammatical construction of the clause, and to ascertain its true reading. Do the words, "to lay and collect taxes, duties, imposts, and excises," constitute a distinct, substantial power; and the words, "to pay debts and provide for the common defence, and general welfare of the United States," constitute another distinct and substantial power? Or are the latter words connected with the former, so as to constitute a qualification upon them? This has been a topic of political controversy; and has furnished abundant materials for popular declamation and alarm. If the former be the true interpretation, then it is obvious, that under colour of the generality of the words to "provide for the common defence and general welfare," the government of the United States is, in reality, a government of general and unlimited powers, notwithstanding the subsequent enumeration of specific powers; if the latter be the true construction, then the power of taxation only is given by the clause, and it is limited to objects of a national character, "for the common defence and the general welfare."

§ 905. The former opinion has been maintained by some minds of great ingenuity, and liberality of views. The latter has been the generally received sense of the nation, and seems supported by reasoning at once solid and impregnable. The reading, therefore, which will be maintained in these commentaries, is that, which makes the latter words a qualification of the former; and this will be best illustrated by supplying the words, which are necessarily to be understood in this interpretation. They will then stand thus: "The congress shall have power to lay and collect taxes, duties, imposts, and excises, *in order* to pay the debts, and to provide for the common defence and general welfare of the United States;" that is, for the purpose of paying the public debts, and providing for the common defence and general welfare of the United States. In this sense, congress has not an unlimited power of taxation; but it is limited to specific objects,--the payment of the public debts, and providing for the common defence and general welfare. A tax, therefore, laid by congress for neither of these objects, would be unconstitutional, as an excess of its legislative authority. In what manner this is to be ascertained, or decided, will be considered hereafter. At present, the interpretation of the words only is before us; and the reasoning, by which that already suggested has been vindicated, will now be reviewed.

§ 906. The constitution was, from its very origin, contemplated to be the frame of a national government, of special and enumerated powers, and not of general and unlimited powers. This is apparent, as will be presently seen, from the history of the proceedings of the convention, which framed it; and it has formed the admitted basis of all legislative and judicial reasoning upon it, ever since it was put into operation, by all, who have been its open friends and advocates, as well as by all, who have been its enemies and opponents. If the clause, "to pay the debts and provide for the common defence and general welfare of the United States," is construed to be an independent and substantive grant of power, it not only renders wholly unimportant and unnecessary the subsequent enumeration of specific powers; but it plainly extends far beyond them, and creates a general authority in congress to pass all laws, which they may deem for the common defence or general welfare. Under such circumstances, the constitution would practically create an unlimited national government. The enumerated powers would tend to embarrassment and confusion; since they would only give rise to doubts, as to the true extent of the general power, or of the enumerated powers.

§ 907. One of the most common maxims of interpretation is, (as has been already stated,) that, as an exception strengthens the force of a law in cases not excepted, so enumeration weakens it in cases not enumerated. But, how could it be applied with success to the interpretation of the constitution of the United States, if the enumerated powers were neither exceptions from, nor additions to, the general power to provide for the common defence and general welfare? To give the enumeration of the specific powers any sensible place or operation in the constitution, it is indispensable to construe them, as not wholly and necessarily embraced in the general power. The common principles of interpretation would seem to instruct us, that the different parts of the same instrument ought to be so expounded, as to give meaning to every part, which will bear it. Shall one part of the same sentence be excluded altogether from a share in the meaning; and shall the more doubtful and indefinite terms be retained in their full extent, and the clear and precise expressions be denied any signification? For what purpose could the enumeration of particular powers be inserted, if these and all others were meant to be included in the preceding general power? Nothing is more natural or common, than first to use a general phrase, and then to qualify it by a recital of particulars. But the idea of an enumeration of particulars, which neither explain, nor qualify the general meaning, and can have no other effect, than to confound and mislead, is an absurdity, which no one ought to charge on the enlightened authors of the constitution. It would be to charge them either with premeditated folly, or premeditated fraud.

§ 908. On the other hand, construing this clause in connexion with, and as a part of the preceding clause, giving the power to lay taxes, it becomes sensible and operative. It becomes a qualification of that clause, and limits the taxing power to objects for the common defence or general welfare. It then contains no grant of any power whatsoever; but it is a mere expression of the ends and purposes to be effected by the preceding power of taxation.

§ 909. An attempt has been sometimes made to treat this clause, as distinct and independent, and yet as having no real significancy *per se,* but (if it may be so said) as a mere prelude to the succeeding enumerated powers. It is not improbable, that this mode of explanation has been suggested by the fact, that in the revised draft of the constitution in the convention the clause was separated from the preceding exactly in the same manner, as every succeeding clause was, viz. by a semicolon, and a break in the paragraph; and that it now stands, in some copies, and it is said, that it stands in the official copy, with a semicolon interposed. But this circumstance will be found of very little weight, when the origin of the clause, and its progress to its present state are traced in the proceedings of the convention. It will then appear, that it was first introduced as an appendage to the power to lay taxes. But there is a fundamental objection to the interpretation thus attempted to be maintained, which is, that it robs the clause of all efficacy and meaning. No person has a right to assume, that any part of the constitution is useless, or is without a meaning; and *a fortiori* no person has a right to rob any part of a meaning, natural and appropriate to the language in the connexion, in which it stands. Now, the words have such a natural and appropriate meaning, as a qualification of the preceding clause to lay taxes. Why, then, should such a meaning be rejected?

§ 910. It is no sufficient answer to say, that the clause ought to be regarded, merely as containing "general terms, explained and limited, by the subjoined specifications, and therefore requiring no critical attention, or studied precaution;" because it is assuming the very point in controversy, to assert, that the clause is connected with any subsequent specifications. It is not said, to "provide for the common defence, and general welfare, *in manner following, viz.,*" which would be the natural expression, to indicate such an intention. But it stands entirely disconnected from every subsequent clause, both in sense and punctuation; and is no more a part of them, than they are of the power to lay taxes. Besides; what suitable application, in such a sense, would there be of the last clause in the enumeration, viz., the clause "to make all laws, necessary and proper for carrying into execution the foregoing powers, &c.?" Surely, this clause is as applicable to the power to lay taxes, as to any other; and no one would dream of its being a mere specification, under the power to provide for the common defence, and general welfare.

§ 911. It has been said [in James Madison's 1800 Report on the Virginia Resolutions] in support of this construction, that in the articles of confederation (art. 8) it is provided, that "all charges of war, and all other expenses, that shall be incurred for the common defence, or general welfare, and allowed by the United States in congress assembled, shall be defrayed out of a common treasury, &c.;" and that "the similarity in the use of these same phrases in these two great federal charters may well be considered, as rendering their meaning less liable to misconstruction; because it will scarcely be said, that in the former they were ever understood to be either a general grant or power, or to authorize the requisition or application of money by the old congress to the common defence and [or] general welfare, except in the cases afterwards enumerated, which explained and limited their meaning; and if such was the limited meaning attached to these phrases in the very instrument revised and remodelled by the present constitution, it can never be supposed, that when copied into this constitution, a different meaning ought to be attached to them." Without stopping to consider, whether the constitution can in any just and critical sense be deemed a revision and remodelling of the confederation, if the argument here stated be of any value, it plainly establishes, that the words ought to be construed, as a qualification or limitation of the power to lay taxes. By the confederation, all expenses incurred for the common defence, or general welfare, are to be defrayed out of a common treasury, to be supplied by requisitions upon the states. Instead of requisitions, the constitution gives the right to the national government directly to lay taxes. So, that the only difference in this view between the two clauses is, as to the mode of obtaining the money, not as to the objects or purposes, to which it is to be applied. If then the constitution were to be construed according to the true bearing of this argument, it would read thus: congress shall have power to lay taxes for "all charges of war, and all other expenses, that shall be incurred for the common defence or general welfare." This plainly makes it a qualification of the taxing power; and not an independent provision, or a general index to the succeeding specifications of power. There is not, however, any solid ground, upon which it can be for a moment maintained, that the language of the constitution is to be enlarged, or restricted by the language of the confederation. That would be to make it speak, what its words do not import, and its objects do not justify. It would be to append it, as a codicil, to an instrument, which it was designed wholly to supercede and vacate.

§ 912. But the argument in its other branch rests on an assumed basis, which is not admitted. It supposes, that in the confederation no expenses, not strictly incurred under some of the subsequent specified powers given to the continental congress, could be properly payable out of the common treasury. Now, that is a proposition to be proved; and is not to be taken for granted. The confederation was not finally ratified, so as to become a binding instrument on any of the states, until March, 1781. Until that period there could be no practice or construction under it; and it is not shown, that subsequently there was any exposition to the effect now insisted on. Indeed, after the peace of 1783, if there had been any such exposition, and it had been unfavourable to the broad exercise of the power, it would have been entitled to less weight, than usually belongs to the proceedings of public bodies in the administration of their powers; since the decline and fall of the confederation was so obvious, that it was of little use to exert them. The states notoriously disregarded the rights and prerogatives admitted to belong to the confederacy; and even the requisitions of congress, for objects most unquestionably within their constitutional authority, were openly denied, or silently evaded.

Under such circumstances, congress would have little inclination to look closely to their powers; since, whether great or small, large or narrow, they were of little practical value, and of no practical cogency.

§ 913. But it does so happen, that in point of fact, no such unfavourable or restrictive interpretation or practice was ever adopted by the continental congress. On the contrary, they construed their power on the subject of requisitions and taxation, exactly as it is now contended for, as a power to make requisitions on the states for all expenses, which they might deem proper to incur for the common defence and general welfare; and to appropriate all monies in the treasury to the like purposes. This is admitted to be of such notoriety, as to require no proof. Surely, the practice of that body in questions of this nature must be of far higher value, than the mere private interpretation of any persons in the present times, however respectable. But the practice was conformable to the constitutional authority of congress under the confederation. The ninth article expressly delegates to congress the power "to ascertain the necessary sums to be raised for the service of the United States, and to appropriate and apply the same for defraying the public expenses;" and then provides, that congress shall not "ascertain the sums and expenses necessary for the *defence* and *welfare* of the United States, *or any of them,* &c. unless nine states assent to the same." So that here we have, in the eighth article, a declaration, that "all charges of war and all other expenses, that shall be incurred for the *common defence* or *general* welfare, &c. shall be defrayed out of a common treasury;" and in the ninth article, an express power to ascertain the necessary sums of money to be raised for the public service; and then, that the necessary sums for the defence and welfare of the United States, (and not of the United States alone, for the words are added,) *or of any of them,* shall be ascertained by the assent of nine states. Clearly therefore, upon the plain language of the articles, the words "common defence and general welfare," in one, and "defence and welfare," in another, and "public service," in another, were not idle words, but were descriptive of the very intent and objects of the power; and not confined even to the defence and welfare of all the states, but extending to the welfare and defence of *any of them.* The power then is, in this view, even larger, than that conferred by the constitution.

§ 914. But there is no ground whatsoever, which authorizes any resort to the confederation, to interpret the power of taxation, which is conferred on congress by the constitution. The clause has no reference whatsoever to the confederation; nor indeed to any other clause of the constitution. It is, on its face, a distinct, substantive, and independent power. Who, then, is at liberty to say, that it is to be limited by other clauses, rather than they to be enlarged by it; since there is no avowed connexion, or reference from the one to the others? Interpretation would here desert its proper office, that, which requires, that "every part of the expression ought, if possible, to be allowed some meaning, and be made to conspire to some common end."

§ 915. It has been farther said [in Madison's 1800 Report], in support of the construction now under consideration, that "whether the phrases in question are construed to authorize every measure relating to the common defence and general welfare, as contended by some; or every measure only, in which there might be an application of money, as suggested by the caution of others; the effect must substantially be the same, in destroying the import and force of the particular enumeration of powers, which follow these general phrases in the constitution. For it is evident, that there is not a single power whatsoever, which may not have some reference to the common defence, or the general welfare; nor a power of any magnitude, which, in its exercise, does not involve, or admit an application of money. The government, therefore, which possesses power in either one, or the other of these extents, is a government without limitations, formed by a particular enumeration of powers; and consequently the meaning and effect of this particular enumeration is destroyed by the exposition given to these general phrases." The conclusion deduced from these premises is, that under the confederation, and the constitution, "congress is authorized to provide money for the common defence and general welfare. In both is subjoined to this authority an enumeration of the cases, to which their powers shall extend. Money cannot be applied to the general welfare otherwise, than by an application of it to some particular measure, conducive to the general welfare. Whenever, therefore, money has been raised by the general authority, and is to be applied to a particular measure, a question arises, whether the particular measure be within the enumerated authorities vested in the congress. If it be, the money requisite for it may be applied to it; if it be not, no such application can be made. This fair and obvious interpretation coincides with, and is enforced by the clause in the constitution, which declares, that no money shall be drawn from the treasury but in consequence of appropriations by law. An appropriation of money to the general welfare would be deemed rather a mockery, than an observance of this constitutional injunction."

§ 916. Stripped of the ingenious texture, by which this argument is disguised, it is neither more nor less, than an attempt to obliterate from the constitution the whole clause, "to pay the debts, and provide for the common defence and general welfare of the United States," as entirely senseless, or inexpressive of any intention whatsoever. Strike them out, and the constitution is exactly what the argument contends for. It is, therefore, an argument, that the words ought not to be in the constitution; because if they are, and have any meaning, they enlarge it beyond the scope of certain other enumerated powers, and this is both mischievous and dangerous. Being in the constitution, they are to be deemed, *vox et preterea nihil,* an empty sound and vain phraseology, a finger-board pointing to other powers, but having no use whatsoever, since these powers are sufficiently apparent without. Now, it is not too much to say, that in a constitution of government, framed and adopted by the people, it is a most unjustifiable latitude of interpretation to deny effect to any clause, if it is sensible in the language, in which it is expressed, and in the place, in which it stands. If words are inserted, we are bound to presume, that they have some definite object, and intent; and to reason them out of the constitution upon arguments *ab inconvenienti,* (which to one mind may appear wholly unfounded, and to another wholly satisfactory,) is to make a new constitution, not to construe the old one. It is to do the very thing, which is so often complained of, to make a constitution to suit our own notions and wishes, and not to administer, or construe that, which the people have given to the country.

§ 917. But what is the argument, when it is thoroughly sifted? It reasons upon a supposed dilemma, upon which it suspends the advocates of the two contrasted opinions. If the power to provide for the common defence and general welfare is an independent power, then it is said, that the government is unlimited, and the subsequent enumeration of powers is unnecessary and useless. If it is a mere appendage or qualification of the power to lay taxes, still it involves a power of general appropriation of the monies so raised, which indirectly produces the same result. Now, the former position may be safely admitted to be true by those, who do not deem it an independent power; but the latter position is not a just conclusion from the premises, which it states, that it is a qualified power. It is not a logical, or a practical sequence from the premises; it is a *non sequitur.*

§ 918. A dilemma, of a very different sort, might be fairly put to those, who contend for the doctrine, that the words are not a qualification of the power to lay taxes, and, indeed, have no meaning, or use *per se.* The words are found in the clause respecting taxation, and as a part of that clause. If the power to tax extends simply to the payment of the debts of the United States, then congress has no power to lay any taxes for any other purpose. If so, then congress could not appropriate the money raised to any other purposes; since the restriction is to taxes for payment of the debts of the United States, that is, of the debts *then existing*. This would be almost absurd. If, on the other hand, congress have a right to lay taxes, and appropriate the money to any other objects, it must be, because the words, "to provide for the common defence and general welfare," authorize it, by enlarging the power to those objects; for there are no other words, which belong to the clause. All the other powers are in distinct clauses, and do not touch taxation. No advocate for the doctrine of a restrictive power will contend, that the power to lay taxes to pay *debts,* authorizes the payment of all debts, which the United States may choose to incur, whether for national or constitutional objects, or not. The words, "to pay debts," are therefore, either antecedent debts, or debts to be incurred "for the common defence and general welfare," which will justify congress in incurring any debts for such purposes. But the language is not confined to the payment of debts for the common defence and general welfare. It is not "to pay the debts" merely; but "to *provide for* the common defence and general welfare." That is, congress may lay taxes to provide means for the common defence and general welfare. So that there is a difficulty in rejecting one part of the qualifying clause, without rejecting the whole, or enlarging the words for some purposes, and restricting them for others.

§ 919. A power to lay taxes for any purposes whatsoever is a general power; a power to lay taxes for certain specified purposes is a limited power. A power to lay taxes for the common defence and general welfare of the United States is not in common sense a general power. It is limited to those objects. It cannot constitutionally transcend them. If the defence proposed by a tax be not the common defence of the United States, if the welfare be not general, but special, or local, as contradistinguished from national, it is not within the scope of the constitution. If the tax be not proposed for the common defence, or general welfare, but for other objects, wholly extraneous, (as for instance, for propagating Mahometanism among the Turks, or giving aids and subsidies to a foreign nation, to build palaces for its kings, or erect monuments to its heroes,) it would be wholly indefensible upon constitutional principles. The power, then, is, under such circumstances, necessarily a qualified power. If it is so, how then does it affect, or in the slightest degree trench upon the other enumerated powers? No one will pretend, that the power to lay taxes would, in general, have superseded, or rendered unnecessary all the other enumerated powers. It would neither enlarge, nor qualify them. A power to tax does not include them. Nor would they, (as unhappily the confederation too clearly demonstrated,) necessarily include a power to tax. Each has its appropriate office and objects; each may exist without necessarily interfering with, or annihilating the other. No one will pretend, that the power to lay a tax necessarily includes the power to declare war, to pass naturalization and bankrupt laws, to coin money, to establish post-offices, or to define piracies and felonies on the high seas. Nor would either of these be deemed necessarily to include the power to tax. It might be convenient; but it would not be absolutely indispensable.

§ 920. The whole of the elaborate reasoning upon the propriety of granting the power of taxation, pressed with so much ability and earnestness, both in and out of the convention, as vital to the operations of the national government, would have been useless, and almost absurd, if the power was included in the subsequently enumerated powers. If the power of taxing was to be granted, why should it not be qualified according to the intention of the framers of the constitution? But then, it is said, if congress may lay taxes for the common defence and general welfare, the money may be appropriated for those purposes, although not within the scope of the other enumerated powers. Certainly it may be so appropriated; for if congress is authorized to lay taxes for such purposes, it would be strange, if, when raised, the money could not be applied to them. That would be to give a power for a certain end, and then deny the end intended by the power. It is added, "that there is not a single power whatsoever, which may not have some reference to the common defence or general welfare; nor a power of any magnitude, which, in its exercise, does not involve, or admit an application of money." If by the former language is meant, that there is not any power belonging, or incident to any government, which has not some reference to the common defence or general welfare, the proposition may be peremptorily denied. Many governments possess powers, which have no application to either of these objects in a just sense; and some possess powers repugnant to both. If it is meant, that there is no power belonging, or incident to a good government, and especially to a republican government, which may not have some reference to those objects, that proposition may, or may not be true; but it has nothing to do with the present inquiry. The only question is, whether a mere power to lay taxes, and appropriate money for the common defence and general welfare, does include all the other powers of government; or even does include the other enumerated powers (limited as they are) of the national government. No person can answer in the affirmative to either part of the inquiry, who has fully considered the subject. The power of taxation is but one of a multitude of powers belonging to governments; to the state governments, as well as the national government. Would a power to tax authorize a state government to regulate the descent and distribution of estates; to prescribe the form of conveyances; to establish courts of justice for general purposes; to legislate respecting personal rights, or the general dominion of property; or to punish all offences against society? Would it confide to congress the power to grant patent rights for intervention; to provide for counterfeiting the public securities and coin; to constitute judicial tribunals with the powers confided by the third article of the constitution; to declare war, and raise armies and navies, and make regulations for their government; to exercise exclusive legislation in the territories of the United States, or in other ceded places; or to make all laws necessary and proper to carry into effect all the powers given by the constitution? The constitution itself upon its face refutes any such notion. It gives the power to tax, as a substantive power; and gives others, as equally substantive and independent.

§ 921. That the same means may sometimes, or often, be resorted to, to carry into effect the different powers, furnishes no objection; for that is common to all governments. That an appropriation of money may be the usual, or best mode of carrying into effect some of these powers, furnishes no objection; for it is one of the purposes, for which, the argument itself admits, that the power of taxation is given. That it is indispensable for the due exercise of all the powers, may admit of some doubt. The only real question is, whether even admitting the power to lay taxes is appropriate for some of the purposes of other enumerated powers, (for no one will contend, that it will, of itself, reach, or provide for them all,) it is limited to such appropriations, as grow out of the exercise of those powers. In other words, whether it is an incident to those powers, or a substantive power in other cases, which may concern the common defence and the general welfare. If there are no other cases, which concern the common defence and general welfare, except those within the scope of the other enumerated powers, the discussion is merely nominal and frivolous. If there are such cases, who is at liberty to say, that, being for the common defence and general welfare, the constitution did not intend to embrace them? The preamble of the constitution declares one of the objects to be, to provide for the common defence, and to promote the general welfare; and if the power to lay taxes is in express terms given to provide for the common defence and general welfare, what ground can there be to construe the power, short of the object? To say, that it shall be merely auxiliary to other enumerated powers, and not coextensive with its own terms, and its avowed objects? One of the best established rules of interpretation, one, which common sense and reason forbid us to overlook, is, that when the object of a power is clearly defined by its terms, or avowed in the context, it ought to be construed, so as to obtain the object, and not to defeat it. The circumstance, that so construed the power may be abused, is no answer. All powers may be abused; but are they then to be abridged by those, who are to administer them, or denied to have any operation? If the people frame a constitution, the rulers are to obey it. Neither rulers, nor any other functionaries, much less any private persons, have a right to cripple it, because it is according to their own views inconvenient, or dangerous, unwise or impolitic, of narrow limits, or of wide influence.

§ 922. Besides; the argument itself admits, that "congress is authorized to provide money for the common defence and general welfare." It is not pretended, that, when the tax is laid, the specific objects, for which it is laid, are to be specified, or that it is to be solely applied to those objects. That would be to insert a limitation, no where stated in the text. But it is said, that it must be applied to the general welfare; and that can only be by an application of it to some *particular measure,* conducive to the general welfare. This is admitted. But then, it is added, that this particular measure must be within the enumerated authorities vested in congress, (that is, within some of the powers not embraced in the first clause,) otherwise the application is not authorized. Why not, since it is for the general welfare? No reason is assigned, except, that not being within the scope of those enumerated powers, it is not given by the constitution. Now, the premises may be true; but the conclusion does not follow, unless the words *common defence* and *general welfare* are limited to the specifications included in those powers. So, that after all, we are led back to the same reasoning, which construes the words, as having no meaning *per se,* but as dependent upon, and an exponent of, the enumerated powers. Now, this conclusion is not justified by the natural connexion or collocation of the words; and it strips them of all reasonable force and efficacy. And yet we are told, that "this fair and obvious interpretation coincides with, and is enforced by, the clause of the constitution, which provides, that no money shall be drawn from the treasury, but in consequence of appropriations by law;" as if the clause did not equally apply, as a restraint upon drawing money, whichever construction is adopted. Suppose congress to possess the most unlimited power to appropriate money for the general welfare; would it not be still true, that it could not be drawn from the treasury, until an appropriation was made by some law passed by congress? This last clause is a limitation, not upon the powers of congress, but upon the acts of the executive, and other public officers, in regard to the public monies in the treasury.

§ 923. The argument in favour of the construction, which treats the clause, as a qualification of the power to lay taxes, has, perhaps, never been presented in a more concise or forcible shape, than in an official opinion, deliberately given by one of our most distinguished statesmen [Jefferson]. "To lay taxes to provide for the general welfare of the United States, is," says he, "to lay taxes *for the purpose,* of providing for the general welfare. For the laying of taxes is the *power,* and the general welfare the *purpose,* for which the power is to be exercised. Congress are not to lay taxes *ad libitum,* for any purpose they please; but only to pay the debts, or provide for the welfare of the Union. In like manner they are not to do any thing they please, to provide for the general welfare; but only to lay taxes for that purpose. To consider the latter phrase, not as describing the purpose of the first, but as giving a distinct and independent power to do any act they please, which might be for the good of the Union, would render all the preceding and subsequent enumerations of power completely useless. It would reduce the whole instrument to a single phrase, that of instituting a congress with power to do whatever would be for the good of the United States; and, as they would be the sole judges of the good or evil, it would also be a power to do whatever evil they pleased. It is an established rule of construction, where a phrase will bear either of two meanings, to give that, which will allow some meaning to the other parts of the instrument, and not that, which will render all the others useless. Certainly, no such universal power was meant to be given them. It was intended to lace them up strictly within the enumerated powers, and those, without which, as means, those powers could not be carried into effect."

§ 924. The same opinion has been maintained at different and distant times by many eminent statesmen. It was avowed, and apparently acquiesed in, in the state conventions, called to ratify the constitution: and it has been, on various occasions, adopted by congress, and may fairly be deemed, that which the deliberate sense of a majority of the nation has at all times supported. This, too, seems to be the construction maintained by the Supreme Court of the United States. In the case of *Gibbons* v. *Ogden,* Mr. Chief Justice Marshall, in delivering the opinion of the court, said, "Congress is authorized to lay and collect taxes, & c. to pay the debts, and provide for the common defence and general welfare of the United States. This does not interefere with the power of the states to tax for the support of their own governments; nor is the exercise of that power by the states an exercise of any portion of the power, that is granted to the United States. In imposing taxes for state purposes, they are not doing, what congress is empowered to do. Congress is not empowered to tax for those purposes, which are within the exclusive province of the states. When, then, each government is exercising the power of taxation, neither is exercising the power of the other." Under such circumstances, it is not, perhaps, too much to contend, that it is the truest, the safest, and the most authoritative construction of the constitution.

Source: *The Founders' Constitution* Volume 2, Article 1, Section 8, Clause 1, Document 28http://press-pubs.uchicago.edu/founders/documents/a1_ _8_1s28.html The University of Chicago Press.

Source: Story, Joseph. *Commentaries on the Constitution of the United States.* 3 vols. Boston, 1833. Vol. 2, 460-464. Published 2000 http://press-pubs.uchicago.edu/founders.

[S]wamp and overflowed lands ... which may be or are found unfit for cultivation

---

SWAMP LAND ACTS OF 1849-1850

Introduction

Early in American history, the U.S. government used the sale of land not only to raise money but also to encourage westward expansion. Not surprisingly, swamp land did not sell particularly well. Moreover, swamps often hindered transportation, seemed to promote flooding, and significantly reduced arable land, especially in states like Florida and Louisiana, where it was so prevalent.

Beginning with this law that it passed in 1849, which applied specifically to Louisiana, Congress granted these lands to the states in which they were found in the hopes that they would drain the swamps and wetlands to encourage settlement. By 1906, states had claimed over 82 million acres under these acts of legislation.

Because the line between reclamation and the creation of other hazards can be, this decision often resulted in the loss of local flora and fauna, salt-water intrusion, and increased risks of flooding, especially in Florida's Everglades, where former swamp lands did not always prove to be as productive for farmlands as was anticipated. Governmental actions further blunted the distinction between natural and man-made disasters (See Hokanson, 2016).

In the twentieth and twenty-first centuries, states and businesses would later criticize the intrusiveness of federal laws that sought to protect wetlands by seeking to preserve such wetlands against drainage and development.

Be it enacted by the Senate and House of Representatives of the United States of America in Congress assembled, That to aid the State of Louisiana in constructing the necessary levees and drains to reclaim the swamp and overflowed lands therein the whole of these swamp and overflowed lands which may be or are found unfit for cultivation shall be, and the same are hereby granted to that State.

Source: Excerpt taken from J.O. Wright,1907, *Swamp and Overflowed Lands in the United States. Ownership and Reclamation.* Washington: Government Printing Office for U.S. Department of Agriculture, p. 5.

PART III

The Civil War and Reconstruction

All persons born or naturalized in the United States and subject to the jurisdiction thereof, are citizens of the United States and of the State wherein they reside

---

THE POST BELLUM AMENDMENTS

1865-1868

Introduction

When 13 southern states attempted to secede in 1861, Republican President Abraham Lincoln thought it was his duty to preserve the Union. The nation entered its bloodiest war, which lasted until the Union victory in 1865. Much of the division between the North and the South had centered on the fact that northern states had outlawed slavery and southern states had not. By war's end, it was clear that the nation could not thus remain half slave and half free.

The nation accordingly adopted three amendments. The first outlawed slavery. The second sought to define citizenship so as to include African-Americans, whose national citizenship had been denied in the Dred Scott Decision of 1857. It further guaranteed all citizens certain fundamental rights. The final amendment prohibited discrimination in voting based on "race, color, or previous condition of servitude."

Although these amendments did not eliminate federalism, and the dual citizenship that most citizens thereby shared, it imposed greater responsibility upon Congress for assuring rights. In time, the Supreme Court would use the Fourteenth Amendment to declare that most provisions of the Bill of Rights, which the Court had previously applied only as limits on the national government, would also limit the states.

The size of the national government grew exponentially during the Civil War. The widened understanding of national citizenship provided justification for increased federal efforts to aid Freedmen, many of whom were displaced from their plantations, after the war. In time, this and other actions also played a role in the assumption of greater federal responsibilities over other areas of life, including disaster relief.

Thirteenth Amendment

Section 1. Neither slavery nor involuntary servitude, except as a punishment for crime whereof the party shall have been duly convicted, shall exist within the United States, or any place subject to their jurisdiction.

Section 2. Congress shall have power to enforce this article by appropriate legislation.

Fourteenth Amendment

Section 1. All persons born or naturalized in the United States and subject to the jurisdiction thereof, are citizens of the United States and of the State wherein they reside. No State shall make or enforce any law which shall abridge the privileges or immunities of citizens of the United States; nor shall any State deprive any person of life, liberty, or property, without due process of law; nor deny to any person within its jurisdiction the equal protection of the laws.

Sections 2-5 omitted.

Fifteenth Amendment

Section 1. The right of citizens of the United States to vote shall not be denied or abridged by the United States or by any State on account of race, color, or previous condition of servitude

A bureau of refugees, freedmen, and abandoned lands

LAW CREATING THE FREEDMEN'S BUREAU

March 3, 1865

Introduction

Nothing shows the interrelationship between man-made and natural disasters like the problems the government faced at the end of the Civil War. As with natural disasters, the war had killed and displaced thousands of refugees, who were in need of food and other provisions, and the Emancipation Proclamation, soon to be followed by the Thirteenth Amendment, had freed former slaves. Many of these freedmen understandably wanted to leave the plantations where they had previously been enslaved. The establishment of the Freedman's Bureau was one of the most aggressive social welfare measures that the national government had ever undertaken in either war or in peace. Despite Section 5 of the law, the national government confiscated relatively little property that had previously belonged to southern slave owners.

Many of the initiatives outlined here would continue through the period of congressional reconstruction of the South (1865-1877). The legacy of slavery continues to hang over race relations with African-Americans still often earning far less than their white counterparts.

CHAP. XC.–

An Act to establish a Bureau for the Relief of Freedmen and Refugees.

Be it enacted by the Senate and House of Representatives of the United States of America in Congress assembled, That there is hereby established in the War Department, to continue during the present war of rebellion, and for one year thereafter, a bureau of refugees, freedmen, and abandoned lands, to which shall be committed, as hereinafter provided, the supervision and management of all abandoned lands, and the control of all subjects relating to refugees and freedmen from rebel states, or from any district of country within the territory embraced in the operations of the army, under such rules and regulations as may be prescribed by the head of the bureau and approved by the President. The said bureau shall be under the management and control of a commissioner to be appointed by the President, by and with the advice and consent of the Senate, whose compensation shall be three thousand dollars per annum, and such number of clerks as may be assigned to him by the Secretary of War, not exceeding one chief clerk, two of the fourth class, two of the third class, and five of the first class. And the commissioner and all persons appointed under this act, shall, before entering upon their duties, take the oath of office prescribed in an act entitled "An act to prescribe an oath of office, and for other purposes," approved July second, eighteen hundred and sixty-two, and the commissioner and the chief clerk shall, before entering upon their duties, give bonds to the treasurer of the United States, the former in the sum of fifty thousand dollars, and the latter in the sum of ten thousand dollars, conditioned for the faithful discharge of their duties respectively, with securities to be approved as sufficient by the Attorney-General, which bonds shall be filed in the office of the first comptroller of the treasury, to be by him put in suit for the benefit of any injured party upon any breach of the conditions thereof.

SEC. 2. *And be it further enacted,* That the Secretary of War may direct such issues of provisions, clothing, and fuel, as he may deem needful for the immediate and temporary shelter and supply of destitute and suffering refugees and freedmen and their wives and children, under such rules and regulations as he may direct.

SEC. 3. *And be it further enacted,* That the President may, by and with the advice and consent of the Senate, appoint an assistant commissioner for each of the states declared to be in insurrection, not exceeding ten in number, who shall, under the direction of the commissioner, aid in the execution of the provisions of this act; and he shall give a bond to the Treasurer of the United States, in the sum of twenty thousand dollars, in the form and manner prescribed in the first section of this act. Each of said commissioners shall receive an annual salary of two thousand five hundred dollars in full compensation for all his services. And any military officer may be detailed and assigned to duty under this act without increase of pay or allowances. The commissioner shall, before the commencement of each regular session of congress, make full report of his proceedings with exhibits of the state of his accounts to the President, who shall communicate the same to congress, and shall also make special reports whenever required to do so by the President or either house of congress; and the assistant commissioners shall make quarterly reports of their proceedings to the commissioner, and also such other special reports as from time to time may be required.

SEC. 4. *And be it further enacted,* That the commissioner, under the direction of the President, shall have authority to set apart, for the use of loyal refugees and freedmen, such tracts of land within the insurrectionary states as shall have been abandoned, or to which the United States shall have acquired title by confiscation or sale, or otherwise, and to every male citizen, whether refugee or freedman, as aforesaid, there shall be assigned not more than forty acres of such land, and the person to whom it was so assigned shall be protected in the use and enjoyment of the land for the term of three years at an annual rent not exceeding six per centum upon the value of such land, as it was appraised by the state authorities in the year eighteen hundred and sixty, for the purpose of taxation, and in case no such appraisal can be found, then the rental shall be based upon the estimated value of the land in said year, to be ascertained in such manner as the commissioner may by regulation prescribe. At the end of said term, or at any time during said term, the occupants of any parcels so assigned may purchase the land and receive such title thereto as the United States can convey, upon paying therefor the value of the land, as ascertained and fixed for the purpose of determining the annual rent aforesaid.

SEC. 5. *And be it further enacted,* That all acts and parts of acts inconsistent with the provisions of this act, are hereby repealed.

APPROVED, March 3, 1865.

Source: U.S., *Statutes at Large, Treaties, and Proclamations of the United States of America,* vol. 13 (Boston, 1866), pp. 507–9.

Source: Freedmen & Southern Society Project. http://www.freedmen.umd.edu/fbact.htm. 12/14/2015.

They can attain to that condition only through their own merits and exertions

---

PRESIDENT ANDREW JOHNSON'S VETO OF FREEDMAN'S BUREAU ACT

(1866)

Introduction

The line between man-made and natural disasters is often fine and intertwined. Even before Abraham Lincoln issued the Emancipation Proclamation on January 1, 1863, slaves were seeking their freedom by moving to Union encampments. Once the Thirteenth Amendment (1865) abolished involuntary servitude, many slaves naturally sought to leave the plantations where masters had held them in bondage. They also sought schools for their children and other social services. Many of these services, like later disaster aid, were implemented by the Freedmen's Bureau, previously called the Bureau of Refugees, Freedmen and Abandoned Lands, which was established in 1865.

In this veto of an amendment to the bill in 1866, President Andrew Johnson expressed his reservations about the law, which would be reenacted in an amended form, and which continued into the administration of Ulysses S. Grant.

To the Senate of the United States:

I have examined with care the bill which originated in the Senate, and has been passed by the two Houses of Congress, to amend an act entitled "An act to establish a bureau for the relief of freedmen and refugees, and for other purposes." Having, with much regret, come to the conclusion that it would not be consistent with the public welfare to give my approval to the measure, I return the bill to the Senate with my objections to its becoming a law.

I might call to mind, in advance of these objections, that there is no immediate necessity for the proposed measure. The act to establish a bureau for the relief of freedmen and refugees, which was approved in the month of March last, has not yet expired. It was thought stringent and extensive enough for the purpose in view. Before it ceases to have effect, further experience may assist to guide us to a wise conclusion as to the policy to be adopted in time of peace.

I have, with Congress, the strongest desire to secure to the freedmen the full enjoyment of their freedom and their property and their entire independence and equality in making contracts for their labor. But the bill before me contains provisions which, in my opinion, are not warranted by the Constitution and are not well suited to accomplish the end in view. The bill proposes to establish by authority of Congress military jurisdiction over all parts of the United States containing refugees and freedmen. It would, by its very nature, apply with the most force to those parts of the United States in which the freedmen most abound; and it expressly extends the existing temporary jurisdiction of the Freedmen's Bureau, with greatly enlarged powers, over those States in which the ordinary course of judicial proceedings has been interrupted by the rebellion. The source from which this military jurisdiction is to emanate is none other than the President of the United States, acting through the War Department and the Commissioner of the Freedmen's Bureau. The agents to carry out this military jurisdiction are to be selected either from the army or from civil life. The country is to be divided into districts and sub-districts, and the number of salaried agents to be employed may be equal to the number of counties or parishes in all the United States where freedmen and refugees are to be found. The subjects over which this military jurisdiction is to extend in every part of the United States, include protection to all employees, agents, and officers of this bureau in the exercise of the duties imposed upon them by the bill. In eleven States it is further to extend over all cases affecting freedmen and refugees discriminated against by local law, custom, or prejudice. In those eleven States the bill subjects any white person who may be charged with depriving a freedman of any civil rights or immunities belonging to white persons to imprisonment or fine, or both, without, however, defining the civil rights and immunities which are thus to be secured to the freedmen by military law. This military jurisdiction also extends to all questions that may arise respecting contracts....

The third section of the bill authorizes a general and unlimited grant of support to the destitute and suffering refugees and freedmen, and their wives and children. Succeeding sections make provision for the rent or purchase of landed estates for freedmen, and for the erection, for their benefit, of suitable asylums and schools, the expenses to be defrayed from the treasury of the whole people. The Congress of the United States has never, heretofore, thought itself competent to establish asylums beyond the limits of the District of Columbia, except for the benefit of our disabled soldiers and sailors. It has never founded schools for any class of our own people, not even for the orphans of those who have fallen in the defence of the Union, but has left the care of their education to the much more competent and efficient control of the States, of communities, of private associations, and of individuals. It has never deemed itself authorized to expend the public money for the rent or purchase of homes for the thousands, not to say millions, of the white race who are honestly toiling from day to day for their subsistence. A system for the support of indigent persons in the United States was never contemplated by the authors of the Constitution. Nor can any good reason be advanced why, as a permanent establishment, it should be founded for one class or color of our people more than for another. Pending the war, many refugees and freedmen received support from the Government, but it was never intended that they should henceforth be fed, clothed, educated, and sheltered by the United States. The idea on which the slaves were assisted to freedom was, that on becoming free they would be a self-sustaining population. Any legislation that shall imply that they are not expected to attain a self-sustaining condition must have a tendency injurious alike to their character and their prosperity. The appointment of an agent for every county and parish will create an immense patronage, and the expense of the numerous officers and their clerks to be appointed by the President will be great in the beginning, with a tendency steadily to increase. The appropriations asked by the Freedmen's Bureau, as now established, for the year 1866, amount to $11,745,000. It may be safely estimated that the cost to be incurred under the pending bill will require double that amount, more than the entire sum expended in any one year under the administration of the second Adams. If the presence of agents in every parish and county is to be considered as a war measure, opposition or even resistance might be provoked, so that to give effect to their jurisdiction troops would have to be stationed within reach of every one of them, and thus a standing army be rendered necessary. Large appropriations would therefore be required to sustain and enforce military jurisdiction in every county or parish from the Potomac to the Rio Grande. The condition of our fiscal affairs is encouraging; but, in order to sustain the present measure of public confidence, it is necessary that we practice not merely customary economy, but, as far as possible, severe retrenchment....

There are still further objections to the bill, on grounds seriously affecting the class of persons to whom it is designed to bring relief. It will tend to keep the mind of the freedman in a state of uncertain expectation and restlessness, while to those among whom he lives it will be a source of constant and vague apprehension. Undoubtedly the freedman should be protected by the civil authorities, especially by the exercise of all the constitutional powers of the courts of the United States and of the States. His condition is not so exposed as may at first be imagined. He is in a portion of the country where his labor cannot well be spared. Competition for his services from planters, from those who are constructing or repairing railroads, or from capitalists in his vicinage, or from other States, will enable him to command almost his own terms. He also possesses a perfect right to change his place of abode, and if, therefore, he does not find in one community or State a mode of life suited to his desires, or proper remuneration for his labor, he can move to another where labor is more esteemed and better rewarded. In truth, however, each State, induced by its own wants and interests, will do what is necessary and proper to retain within its borders all the labor that is needed for the development of its resources. The laws that regulate supply and demand will maintain their force, and the wages of the laborer will be regulated thereby. There is no danger that the great demand for labor will not operate in favor of the laborer. Neither is sufficient consideration given to the ability of the freedmen to protect and take care of themselves. It is no more than justice to them to believe that, as they have received their freedom with moderation and forbearance, so they will distinguish themselves by their industry and thrift, and soon show the world that in a condition of freedom they are self-sustaining and capable of selecting their own employment and their own places of abode; of insisting for themselves on a proper remuneration, and of establishing and maintaining their own asylums and schools. It is earnestly hoped that instead of wasting away, they will, by their own efforts, establish for themselves a condition of respectability and prosperity. It is certain that they can attain to that condition only through their own merits and exertions. In this connection the query presents itself, whether the system proposed by the bill will not, when put into complete operation, practically transfer the entire care, support, and control of four millions of emancipated slaves to agents, overseers, or taskmasters, who, appointed at Washington, are to be located in every county and parish throughout the United States containing freedmen and refugees. Such a system would inevitably tend to such a concentration of power in the Executive as would enable him, if so disposed, to control the action of a numerous class, and to use them for the attainment of his own political ends.... [Johnson expresses further concern that Congress had not seated delegates from the southern states.]

I return the bill to the Senate in the earnest hope that a measure involving questions and interests so important to the country will not become a law, unless upon deliberate consideration by the people it shall receive the sanction of an enlightened public judgment.

Andrew Johnson
Washington, D.C., Feb. 19, 1866

Source: *Andrew Johnson, His Life and Speeches* by Lillian Foster, New York: Richardson & Co., 1866.

Congress has no power to appropriate this money as an act of charity

---

Attributed to David Crockett
HARPER'S MAGAZINE

January, 1867, p. 67

Introduction

Few individuals have achieved the place in American folklore as David (often called Davy) Crockett (1786-1836), a Tennessee-born member of the U.S. House of Representatives who considered a run for the presidency. He is chiefly known for his home-spun witticisms, his heroic claims about battling wild beasts, and his colorful language.

This anecdote was published in the January 1867 issue of *Harper's Magazine* and was written by Edward S. Ellis (1849-1916) under the pen name James J. Bethune. It claims that Crockett refused to vote for a public appropriation for the benefit of a widow who had lost her husband at sea. The longer story goes on to tell how Crockett had been upbraided by a constituent, Horatio Bunce, in response to Crockett's putative earlier vote for aid for relief of victims of a fire.

Although Crockett does appear to have voted against the appropriation for the widow, he does not appear to have given a speech, and Congress does appear to have approved of it. Now regarded as a fabrication,[1] the story is widely believed and continues to express concern that governmental relief efforts cannot be constitutionally justified.

---

[1] See AnnTtoplovich and Bruce E. Maker (atop@tennesseehistory.org), "'Not Yours to Give' Update," http:/h-net.msu.edu/cgi-bin/logbrowse.pl?trx=vx&lisy=H-South&month=0909&week=e&... 10/4/2016.

One day in the House of Representatives, a bill was taken up appropriating money for the benefit of a widow of a distinguished naval officer. Several beautiful speeches had been made in its support. The Speaker was just about to put the question when Davy Crockett arose:

> "Mr. Speaker–I have as much respect for the memory of the deceased, and as much sympathy for the sufferings of the living, if suffering there be, as any man in this House, but we must not permit our respect for the dead or our sympathy for a part of the living to lead us into an act of injustice to the balance of the living. I will not go into an argument to prove that Congress has no power to appropriate this money as an act of charity. Every member upon this floor knows it. We have the right, as individuals, to give away as much of our own money as we please in charity; but as members of Congress we have no right so to appropriate a dollar of the public money. Some eloquent appeals have been made to us upon the ground that it is a debt due the deceased. Mr. Speaker, the deceased lived long after the close of the war; he was in office to the day of his death, and I have never heard that the government was in arrears to him.
>
> Every man in this House knows it is not a debt. We cannot, without the grossest corruption, appropriate this money as the payment of a debt. We have not the semblance of authority to appropriate it as a charity. Mr. Speaker, I have said we have the right to give as much money of our own as we please. I am the poorest man on this floor. I cannot vote for this bill, but I will give one week's pay to the object, and if every member of Congress will do the same, it will amount to more than the bill asks."

He took his seat. Nobody replied. The bill was put upon its passage, and, instead of passing unanimously, as was generally supposed, and as, no doubt, it would, but for that speech, it received but few votes, and, of course, was lost.

Resulting advantage to the public does not of itself give to the means by which it is produced the character of a public service

---

LOWELL V. BOSTON

111 Mass. 454 (1873)

Introduction

Although it is commonplace for contemporary governments to present a smorgasbord of programs in the wake of natural disasters, nineteenth century lawyers continued to wrestle with the constitutionality of such aid. In this case arising from Boston, plaintiffs successfully challenged loans that the city was providing through bonds for residence whose homes and livelihoods had been destroyed by a fire the previous year. In this decision, issued by Justice John Wells of the Massachusetts Supreme Judicial Court, the Court attempted to distinguish aid designed for the general welfare from that tailored to private benefit and decided that the law unconstitutionally promoted the latter.

WELLS, J. This is a proceeding under the provisions of the Gen. Sts. c. 18, § 79, to restrain the city of Boston from issuing its bonds for the purpose of raising a fund to be appropriated to the object of rendering aid, by way of loans, in rebuilding upon that portion of the city which was burned over in November 1872. The issue of bonds for that purpose, to an amount not exceeding $20,000,000, was expressly authorized by the St. of 1872, c. 364. The question, therefore, is distinctly presented whether the authority thus conferred upon the city is contrary to the provisions of the Constitution of the Commonwealth.

The issue of bonds by the city, whatever provision may be made for their redemption, involves the possible and not improbable consequence of a necessity to provide for their payment by the city. The right to incur the obligation implies the right to raise money by taxation for payment of the bonds; or, what is equivalent, the right to levy a tax for the purposes for which the fund is to be raised by means of the bonds so authorized.

It is a question, not of municipal authority, but of legislative power. The point of difficulty is not as to the distribution of the burden by allowing it to be imposed upon a limited district within the State; but as to the right of the Legislature to impose or authorize any tax for the object contemplated by this statute.

The power to levy taxes is founded on the right, duty and responsibility to maintain and administer all the governmental functions of the State, and to provide for the public welfare. To justify any exercise of the power requires that the expenditure which it is intended to meet shall be for some public service, or some object which concerns the public welfare. The promotion of the interests of individuals, either in respect of property or business, although it may result incidentally in the advancement of the public welfare, is, in its essential character, a private and not a public object. However certain and great the resulting good to the general public, it does not, by reason of its comparative importance, cease to be incidental. The incidental advantage to the public, or to the State, which results from the promotion of private interests, and the prosperity of private enterprises or business, does not justify their aid by the use of public money raised by taxation, or for which taxation may become necessary. It is the essential character of the direct object of the expenditure which must determine its validity, as justifying a tax, and not the magnitude of the interests to be affected, nor the degree to which the general advantage of the community, and thus the public welfare, may be ultimately benefited by their promotion.

The principle of this distinction is fundamental. It underlies all government that is based upon reason rather than upon force. It is expressed in various forms in the Constitution of Massachusetts. In Art. XI. of c. 2, § 1, by restricting the issuing of moneys from the treasury to purposes of "the necessary defence and support of the Commonwealth; and for the protection and preservation of the inhabitants thereof, agreeably to the acts and resolves of the General Court." In Art. IV. of c. 1, § 1, by declaring the purposes, for which the power of taxation, in its various forms, may be exercised by the General Court, to be "for the public service, in the necessary defence and support of the government of the said Commonwealth, and the protection and preservation of the subjects thereof." The purport and scope of these provisions are made more distinct, and the essential idea upon which they rest is disclosed by reference to the preceding Declaration of Rights, by which the theory and purpose of this frame of government were set forth by its founders. Art. X. declares, "Each individual of the society has a right to be protected by it in the enjoyment of his life, liberty and property, according to standing laws. He is obliged, consequently, to contribute his share to the expense of this protection; to give his personal service, or an equivalent, when necessary: but no part of the property of any individual can, with justice, be taken from him, or applied to public uses, without his own consent, or that of the representative body of the people. In fine, the people of this Commonwealth are not controllable by any other laws than those to which their constitutional representative body have given their consent. And whenever the public exigencies require that the property of any individual should be appropriated to public uses, he shall receive a reasonable compensation therefor."

The power of the government, thus constituted, to affect the individual in his private rights of property, whether by exacting contributions to the general means, or by sequestration of specific property, is confined, by obvious implication as well as by express terms, to purposes and objects alone which the government was established to promote, to wit, public uses and the public service. This power, when exercised in one form, is taxation; in the other, is designated as the right of eminent domain. The two are diverse in respect of the occasion and mode of exercise, but identical in their source, to wit, the necessities of organized society; and in the end by which alone the exercise of either can be justified, to wit, some public service or use. It is due to their identity in these respects that the two powers, otherwise so unlike, are associated together in the same article. So far as it concerns the question what constitutes public use or service that will justify the exercise of these sovereign powers over private rights of property, which is the main question now to be solved, this identity renders it unnecessary to distinguish between the two forms of exercise, as the same tests must apply to and control in each. An appropriation of money raised by taxation, or of property taken by right of eminent domain, by way of gift to an individual for his own private uses exclusively, would clearly be an excess of legislative power.

The distinction between this and its appropriation for the construction of a highway, is marked and obvious. It is independent of all considerations of resulting advantage. The individual, by reason of his capacity, enterprise or situation, might be enabled to employ the money or property thus conferred upon him in such a manner as to furnish employment to great numbers of the community, to give a needed impulse to business of various kinds, and thus promote the general prosperity and welfare. In this view, it might be shown to be for the public good to take from the unenterprising and thriftless their unemployed capital and entrust it to others who will use it to better advantage for the interests of the community. But it needs no argument to show that such an arbitrary exercise of power would be a violation of the constitutional rights of those from whom the money or property was taken, and an unjustifiable usurpation.

In the case of a highway, on the other hand, its direct purpose of public use determines conclusively the question in support of the exercise, both of the right of eminent domain and of taxation, however trifling the advantage to the public compared with that to individuals. The extent or value of the public use, and the wisdom and propriety of the appropriation, are matters to be determined exclusively by the Legislature, either directly or by its delegated authority. When the power exists it is not within the province of the court to interfere with its exercise, by any inquiry into its expediency.

The two instances, above referred to, illustrate the sense in which the furthering of the public good by promotion of the interests of many individuals, differs from a public service. A public service may or may not be productive, practically, of public advantage. Resulting advantage to the public does not of itself give to the means by which it is produced the character of a public service.

There are, indeed, many cases in which the sovereign power of government is exercised to affect private rights of property in favor of private parties, either individuals or corporations. Most conspicuous among these are turnpikes and railroads; in whose favor this right of eminent domain is frequently exercised. Private rights are thus taken and transferred, not to the State, but to the private corporation; and the compensation to the persons injured, required by the Constitution, is also rendered from the corporation.

Such an appropriation of property is justified, and can only be justified, by the public service thereby secured in the increased facilities for transportation of freight and passengers, of which the whole community may rightfully avail itself. The franchises of the corporation are held charged with this duty and trust for the performance of the public service, for which they were granted. *Commonwealth v. Wilkinson,* 16 Pick. 175. *Same v. Boston & Maine Railroad,* 3 Cush. 25, 45. *Old Colony & Fall River Railroad Co. v. County of Plymouth,* 14 Gray 155, 161.

This right of eminent domain is often allowed to be exercised in favor of private aqueduct companies. Here, too, the public service, intended as the object of the grant of the right, is obvious. And although the interests of the aqueduct company are ordinarily relied upon to secure the proper performance of the service, yet, in case of any failure or abuse, the obligation may doubtless be otherwise enforced. *Lumbard v. Stearns,* 4 Cush. 60.

The mill acts, so called, are often referred to as authorizing the exercise of the right of eminent domain by private parties for their exclusive private benefit. And the language of the court, used arguendo, has been sometimes such as to imply that the growth and prosperity of manufacturing and other industrial enterprises were of such importance to the public welfare, as to justify the exercise of the right of eminent domain in their behalf, as a public use. *Boston & Roxbury Mill Co. v. Newman,* 12 Pick. 467. *Hazen v. Essex Co. 12 Cush.* 475, 478. *Talbot v. Hudson,* 16 Gray 417, 426.

That mills for the sawing of lumber for purposes of building, grinding grain for food, and the manufacture of material for clothing, may be of such necessity to a community, especially in the early settlement of a country, as to make their establishment a provision for a public service, we do not question. It is doubtless within the power of the Legislature to declare the existence of a public exigency for the establishment of a mill, for which the right of eminent domain may properly be exercised; as in the case of the Boston & Roxbury Mill Corporation, and the Salem Mill-dam Corporation. What may be the limits of legislative power in that direction, and whether there are any limits except in the sound discretion of the Legislature, it is needless now to inquire. We are satisfied that the mill acts are not founded upon that power, and do not authorize its exercise.

The advantages to be derived from a running stream by the several riparian proprietors, are of natural right. Each one may make use of its waters, as they flow through his lands, in a reasonable manner, for such purposes as they are adapted to serve. In order that each may have his opportunity in turn, each is entitled to have the water allowed to flow to and from his land as it has been accustomed to flow, with only such modifications as result from such reasonable use. Hence, all proprietors upon a stream, from its source to its mouth, have, in a certain sense, a common interest in it, and a common right to the enjoyment of all its capacities. Among those capacities no one is more important than that of the force of the current to supply power for the operation of mills. To make that force practicably serviceable requires a considerable head and fall at the point where it is to be applied; often more than can be gained within the limits of one proprietor. The use of the stream in this mode has always been regarded as a reasonable use, notwithstanding the effect of the dam, by which the head is created, to retard the water in its flow to the proprietor below, and to set it back and thus diminish or destroy the force of the current above. One who thus appropriates the force of the current is in the enjoyment of a common right, in which he is protected, although he may thereby prevent a like use subsequently by the proprietor above. *Hatch v. Dwight,* 17 Mass. 289, 296. *Cary v. Daniels,* 8 Met. 466. *Gould v. Boston Duck Co.* 13 Gray 442. But this protection extends no farther than to justify the appropriation of a part of that quality of the stream, which, until so appropriated, is common to all. It does not justify any, even the least, injury to land outside the channel. Without some law to control, the mill owner would be exposed, not merely to the liability to make just compensation for injuries thus occasioned, but to harassing suits for damages and to abatement of his dam as causing a nuisance. This liability and the inevitable controversies growing out of conflicting rights in the stream itself, tending to defeat all advantageous use of its power, led to the adoption of laws regulating and protecting the beneficial use of streams for mill purposes. The St. of 1795, c. 74, is introduced by the recital, "Whereas the erection and support of mills, to accommodate the inhabitants of the several parts of the State, ought not to be discouraged by many doubts and disputes, and some special provisions are found necessary relative to flowing adjacent lands and mills held by several proprietors." But there is no public service secured through the mill acts, except so far as it may result incidentally, and as the inducements of private interest may lead mill owners to devote their mills to purposes favorable to the public accommodation. The same rights and protection are secured to all who may be possessed of sites for mills, whatever the purpose for which their mills may be designed, and however useless for all purposes of public accommodation or advantage. There is no discrimination in this respect, and no provision to secure any public service that may be supposed to have been contemplated. Further than this, each proprietor is allowed to avail himself of the rights secured by the mill acts, in his own mode and for his own purposes, at his own discretion, without the intervention of any public officer or other tribunal or board, to whom such a governmental function as the exercise of the right of eminent domain is ordinarily entrusted, when not under the special direction of the Legislature itself.

A consideration, still more conclusive to this point, is, that in fact no private property, or right in the nature of property, is taken by force of the mill acts, either for public or private use....

This regulation of the rights of riparian proprietors, both in respect to the stream and to their adjacent lands, liable to be affected by its use, involves no other governmental power than that "to make, ordain and establish all manner of wholesome and reasonable orders, laws, statutes and ordinances," as the General Court "shall judge to be for the good and welfare of this Commonwealth, and for the government and ordering thereof, and of the subjects of the same." Const. of Mass. c. 1, § 1, Art. IV.

All individual rights of property are held subject to this power, which alone can adjust their manifold relations and conflicting tendencies. The absolute right of the individual must yield to and be modified by corresponding rights in other individuals in the community. The resulting general good of all, or the public welfare, is the foundation upon which the power rests, and in behalf of which it is exercised; whether by restricting the use of private property in a manner prejudicial to the public; *Commonwealth v. Alger*, 7 Cush. 53; or by imposing burdens upon it for the protection or convenience in part of the public; Goddard, Petitioner, 16 Pick. 504. *Baker v. Boston*, 12 Pick. 184, 193. *Salem v. Eastern Railroad Co.* 98 Mass. 431; or by modifying rights of individuals, in respect of their mutual relations, in order to secure their more advantageous enjoyment by each....

Upon the same principle, proprietors of wharves, or of general fields, affected by a common interest or a common necessity, are allowed to adopt measures to secure their common advantage, although burdens or restrictions result therefrom which must be shared by the minority, as well as the majority, by whose determination the measures were adopted. Gen. Sts. c. 67. *Wright v. Boston*, 9 Cush. 233.

No other power was exercised for the construction of drains and sewers, until 1841, when cities and towns were authorized to exercise for that object the power of taxation. St. 1841, c. 115. The property in such drains and sewers was by the same act vested in the city or town; so that there was a public use as well as a public service, for which that power was delegated. The exercise of the right of eminent domain, for the same object, was delegated to the city of Boston, by the St. of 1857, c. 225, § 1, and to all other cities and towns by the St. of 1869, c. 111....

"The good and welfare of this Commonwealth," for which "reasonable orders, laws, statutes and ordinances" may be made, by force of which private rights of property may be affected, is a much broader and less specific ground of exercise of power than "public use" and "public service." The former expresses the ultimate purpose, or result sought to be attained by all forms of exercise of legislative power over property. The latter imply a direct relation between the primary object of an appropriation and the public enjoyment. The circumstances may be such that the use or service intended to be secured will practically affect only a small portion of the inhabitants or lands of the Commonwealth. The essential point is, that it affects them as a community, and not merely as individuals. Cooley Const. Limit. 531. This distinction is indicated, and recognized as vital, in *Talbot v. Hudson,* 16 Gray 417, 423, 425, and it lies at the foundation of the decision in that case. There was a taking of private property, by direct authority of the Legislature, which the court held to have been intended as an exercise of the constitutional power to take private property for public use, rendering compensation. The main question was, whether the relief of an extensive territory of valuable lands, in a thickly settled agricultural region, from the nuisance of flooding by the waters of a stream, caused by a single dam below, constituted such an object of public concern as to justify the exercise of the power by removing the dam. The court recognized the difficulty that, so far as the removal of the dam benefited each land owner, it was a private use which would not justify the exercise of that power. But the obstruction in the stream injuriously affected "so large a territory, situated in different towns, and owned by a great number of persons," as to give it the character of a public nuisance, the removal of which "would seem to come fairly within the scope of legislative action." While we do not assent to the suggestions in that opinion, that the general provisions of law for the regulation of mills and the improvement of meadows are based upon the constitutional power to appropriate private property under the right of eminent domain, we accord fully with the judgment rendered and the general principle upon which it is founded.

The same principle is developed in *Dorgan v. Boston,* 12 Allen 223, and *Dingley v. Boston,* 100 Mass. 544. The public use, in one case, was in the improvement of the public streets; in the other, in the remedy for a great public nuisance requiring extraordinary measures for its removal. In both there was a great improvement in the character and value of the new buildings erected upon the territory affected, and thus a promotion of the general prosperity and public welfare. This benefit to the community was anticipated, and was doubtless one of the influential inducements to the adoption of the statutes giving authority for the improvements. It was not in this general advantage however, that the justification, under the Constitution, for such an exercise of power was found, but in the direct and special public service.

In *Hazen v. Essex Co.* 12 Cush. 475, before referred to, there was a distinct and clear public service declared as the object of the act conferring the power to destroy private property, to wit, the improvement of the navigation of Merrimack River. The case does not rest upon the general benefit from the establishment of mills.

Without such public use or service expressly declared, or implied from the nature of the object of the expenditure, taxation in any form cannot be justified *Lowell v. Oliver,* 8 Allen 247. *Freeland v. Hastings,* 10 Allen 570. *Dorgan v. Boston,* 12 Allen 223, 240. *Merrick v. Amherst,* Ib. 500.

In *Morse v. Stocker,* 1 Allen 150, it was held that an assessment for the expense of a sidewalk in a street which was so by dedication only, and in which no right of way was secured to the party assessed, or to the public by its acceptance or location by the proper authorities, was unconstitutional and void.

There is no public use or public service declared in the statute now under consideration, and we are of opinion that none can be found in the purposes of its provisions. By its terms the proceeds of the bonds, thereby authorized, are to be expended in loans to persons who are or may become owners of land in Boston, "the buildings upon which were burned by the fire in said Boston on the ninth and tenth days of November," 1872. The ultimate end and object of the expenditure, as indicated by the provisions of the statute itself, is "to insure the speedy rebuilding on said land."

The general result may indeed be thus stated collectively, as a single object of attainment; but the fund raised is intended to be appropriated distributively, by separate loans to numerous individuals, each one of which will be independent of any relation to the others, or to any general purpose, except that of aiding individual enterprise in matters of private business. The property thus created will remain exclusively private property, to be devoted to private uses at the discretion of the owners of the land; with no restriction as to the character of the buildings to be erected, or the uses to which they shall be devoted; and with no obligation to render any service or duty to the Commonwealth, or to the city, -- except to repay the loan, -- or to the community at large or any part of it. If it be assumed that the private interests of the owners will lead them to reestablish warehouses, shops, manufactories, and stores; and that the trade and business of the place will be enlarged or revived by means of the facilities thus provided; still these are considerations of private interest, and, if expressly declared to be the aim and purpose of the act, they would not constitute a public object, in a legal sense.

As a judicial question the case is not changed by the magnitude of the calamity which has created the emergency; nor by the greatness of the emergency, or the extent and importance of the interests to be promoted. These are considerations affecting only the propriety and expediency of the expenditure as a legislative question. If the expenditure is, in its nature, such as will justify taxation under any state of circumstances, it belongs to the Legislature exclusively to determine whether it shall be authorized in the particular case; and however slight the emergency, or limited or unimportant the interests to be promoted thereby, the court has no authority to revise the legislative action.

On the other hand, if its nature is such as not to justify taxation in any and all cases in which the Legislature might see fit to give authority therefor, no stress of circumstances affecting the expediency, importance or general desirableness of the measure, and no concurrence of legislative and municipal action, or preponderance of popular favor in any particular case, will supply the element necessary to bring it within the scope of legislative power....

"[T]here can be no lawful tax which is not laid for a public purpose"

LOAN ASSOCIATION V. TOPEKA

(1874) 87 U.S. 655; 22 L. Ed. 455; 1874 U.S. LEXIS 1457; 20 Wall. 655

Introduction

This case, brought by the Citizen's Savings and Loan Association of Cleveland against the city of Topeka, Kansas, questioned the aid that the city sought to provide in the form of bonds to the King Wrought-Iron Bridge Manufacturing and Iron-Works Company to encourage it to operate within the city. The issue involved whether it was proper for governments to use resources, backed up by the state's power of taxation, to aid private companies. This issue is important because similar arguments were being made about the government's ability to use general tax funds to pay for people who had been affected by specific catastrophes rather than to people throughout the nation. The opinion was written by Samuel Freeman Miller (1816-1890); the dissent was penned by Justice Nathan Clifford (1802-1881).

Mr. Justice MILLER delivered the opinion of the court.

Two grounds are taken in the opinion of the circuit judge and in the argument of counsel for defendant, on which it is insisted that the section of the statute of February 29th, 1872, on which the main reliance is placed to issue the bonds, is unconstitutional.

The first of these is, that by section five of article twelve of the constitution of that State it is declared that provision shall be made by general law for the organization of cities, towns, and villages; and their power of taxation, assessment, borrowing money, contracting debts, and loaning their credit, shall be so restricted as to prevent the abuse of such power.

The argument is that the statute in question is void because it authorizes cities and towns to contract debts, and does not contain any restriction on the power so conferred. But whether the statute which confers power to contract debts should always contain some limitation or restriction, or whether a general restriction applicable to all cases should be passed, and whether in the absence of both the grant of power to contract is wholly void, are questions whose solution we prefer to remit to the State courts, as in this case we find ample reason to sustain the demurrer on the second ground on which it is argued by counsel and sustained by the Circuit Court.

That proposition is that the act authorizes the towns and other municipalities to which it applies, by issuing bonds or loaning their credit, to take the property of the citizen under the guise of taxation to pay these bonds, and use it in aid of the enterprises of others which are not of a public character, thus perverting the right of taxation, which can only be exercised for a public use, to the aid of individual interests and personal purposes of profit and gain.

The proposition as thus broadly stated is not new, nor is the question which it raises difficult of solution.

If these municipal corporations, which are in fact subdivisions of the State, and which for many reasons are vested with quasi legislative powers, have a fund or other property out of which they can pay the debts which they contract, without resort to taxation, it may be within the power of the legislature of the State to authorize them to use it in aid of projects strictly private or personal, but which would in a secondary manner contribute to the public good; or where there is property or money vested in a corporation of the kind for a particular use, as public worship or charity, the legislature may pass laws authorizing them to make contracts in reference to this property, and incur debts payable from that source.

But such instances are few and exceptional, and the proposition is a very broad one, that debts contracted by municipal corporations must be paid, if paid at all, out of taxes which they may lawfully levy, and that all contracts creating debts to be paid in future, not limited to payment from some other source, imply an obligation to pay by taxation.

It follows that in this class of cases the right to contract must be limited by the right to tax, and if in the given case no tax can lawfully be levied to pay the debt, the contract itself is void for want of authority to make it.

If this were not so, these corporations could make valid promises, which they have no means of fulfilling, and on which even the legislature that created them can confer no such power. The validity of a contract which can only be fulfilled by a resort to taxation, depends on the power to levy the tax for that purpose.

It is, therefore, to be inferred that when the legislature of the State authorizes a county or city to contract a debt by bond, it intends to authorize it to levy such taxes as are necessary to pay the debt, unless there is in the act itself, or in some general statute, a limitation upon the power of taxation which repels such an inference.

With these remarks and with the reference to the authorities which support them, we assume that unless the legislature of Kansas had the right to authorize the counties and towns in that State to levy taxes to be used in aid of manufacturing enterprises, conducted by individuals, or private corporations, for purposes of gain, the law is void, and the bonds issued under it are also void. We proceed to the inquiry whether such a power exists in the legislature of the State of Kansas.

We have already said the question in not new. The subject of the aid voted to railroads by counties and towns has been brought to the attention of the courts of almost every State in the Union. It has been thoroughly discussed and is still the subject of discussion in those courts. It is quite true that a decided preponderance of authority is to be found in favor of the proposition that the legislatures of the States, unless restricted by some special provisions of their constitutions, may confer upon these municipal bodies the right to take stock in corporations created to build railroads, and to lend their credit to such corporations. Also to levy the necessary taxes on the inhabitants, and on property within their limits subject to general taxation, to enable them to pay the debts thus incurred. But very few of these courts have decided this without a division among the judges of which they were composed, while others have decided against the existence of the power altogether.

In all these cases, however, the decision has turned upon the question whether the taxation by which this aid was afforded to the building of railroads was for a public purpose. Those who came to the conclusion that it was, held the laws for the purpose valid. Those who could not reach that conclusion held them void. In all the controversy this has been the turning-point of the judgments of the courts. And it is safe to say that no court has held debts created in aid of railroad companies, by counties or towns, valid on any other ground than that the purpose for which the taxes were levied was a public use, a purpose or object which it was the right and the duty of State governments to assist by money raised from the people by taxation.

The argument in opposition to this power has been, that railroads built by corporations organized mainly for purposes of gain -- the roads which they built being under their control, and not that of the State -- were private and not public roads, and the tax assessed on the people went to swell the profits of individuals and not to the good of the State, or the benefit of the public, except in a remote and collateral way. On the other hand it was said that roads, canals, bridges, navigable streams, and all other highways had in all times been matter of public concern. That such channels of travel and of the carrying business had always been established, improved, regulated by the State, and that the railroad had not lost this character because constructed by individual enterprise, aggregated into a corporation.

We are not prepared to say that the latter view of it is not the true one, especially as there are other characteristics of a public nature conferred on these corporations, such as the power to obtain right of way, their subjection to the laws which govern common carriers, and the like, which seem to justify the proposition. Of the disastrous consequences which have followed its recognition by the courts and which were predicted when it was first established there can be no doubt.

We have referred to this history of the contest over aid to railroads by taxation, to show that the strongest advocates for the validity of these laws never placed it on the ground of the unlimited power in the State legislature to tax the people, but conceded that where the purpose for which the tax was to be issued could no longer be justly claimed to have this public character, but was purely in aid of private or personal objects, the law authorizing it was beyond the legislative power, and was an unauthorized invasion of private right.

It must be conceded that there are such rights in every free government beyond the control of the State. A government which recognized no such rights, which held the lives, the liberty, and the property of its citizens subject at all times to the absolute disposition and unlimited control of even the most democratic depository of power, is after all but a despotism. It is true it is a despotism of the many, of the majority, if you choose to call it so, but it is none the less a despotism. It may well be doubted if a man is to hold all that he is accustomed to call his own, all in which he has placed his happiness, and the security of which is essential to that happiness, under the unlimited dominion of others, whether it is not wiser that this power should be exercised by one man than by many.

The theory of our governments, State and National, is opposed to the deposit of unlimited power anywhere. The executive, the legislative, and the judicial branches of these governments are all of limited and defined powers.

There are limitations on such power which grow out of the essential nature of all free governments. Implied reservations of individual rights, without which the social compact could not exist, and which are respected by all governments entitled to the name. No court, for instance, would hesitate to declare void a statute which enacted that A. and B. who were husband and wife to each other should be so no longer, but that A. should thereafter be the husband of C., and B. the wife of D. Or which should enact that the homestead now owned by A. should no longer be his, but should henceforth be the property of B.

Of all the powers conferred upon government that of taxation is most liable to abuse. Given a purpose or object for which taxation may be lawfully used and the extent of its exercise is in its very nature unlimited. It is true that express limitation on the amount of tax to be levied or the things to be taxed may be imposed by constitution or statute, but in most instances for which taxes are levied, as the support of government, the prosecution of war, the National defence, any limitation is unsafe. The entire resources of the people should in some instances be at the disposal of the government.

The power to tax is, therefore, the strongest, the most pervading of all the powers of government, reaching directly or indirectly to all classes of the people. It was said by Chief Justice Marshall, in the case of *McCulloch v. The State of Maryland,* that the power to tax is the power to destroy. A striking instance of the truth of the proposition is seen in the fact that the existing tax of ten per cent. imposed by the United States on the circulation of all other banks than the National banks, drove out of existence every State bank of circulation within a year or two after its passage. This power can as readily be employed against one class of individuals and in favor of another, so as to ruin the one class and give unlimited wealth and prosperity to the other, if there is no implied limitation of the uses for which the power may be exercised.

To lay with one hand the power of the government on the property of the citizen, and with the other to bestow it upon favored individuals to aid private enterprises and build up private fortunes, is none the less a robbery because it is done under the forms of law and is called taxation. This is not legislation. It is a decree under legislative forms.

Nor is it taxation. A "tax," says *Webster's Dictionary,* "is a rate or sum of money assessed on the person or property of a citizen by government for the use of the nation or state." "Taxes are burdens or charges imposed by the legislature upon persons or property to raise money for public purposes."

Coulter, J., in *Northern Liberties v. St. John's Church,* says, very forcibly, "I think the common mind has everywhere taken in the understanding that taxes are a public imposition, levied by authority of the government for the purpose of carrying on the government in all its machinery and operations -- that they are imposed for a public purpose."

We have established, we think, beyond cavil that there can be no lawful tax which is not laid for a public purpose. It may not be easy to draw the line in all cases so as to decide what is a public purpose in this sense and what is not.

It is undoubtedly the duty of the legislature which imposes or authorizes municipalities to impose a tax to see that it is not to be used for purposes of private interest instead of a public use, and the courts can only be justified in interposing when a violation of this principle is clear and the reason for interference cogent. And in deciding whether, in the given case, the object for which the taxes are assessed falls upon the one side or the other of this line, they must be governed mainly by the course and usage of the government, the objects for which taxes have been customarily and by long course of legislation levied, what objects or purposes have been considered necessary to the support and for the proper use of the government, whether State or municipal. Whatever lawfully pertains to this and is sanctioned by time and the acquiescence of the people may well be held to belong to the public use, and proper for the maintenance of good government, though this may not be the only criterion of rightful taxation.

But in the case before us, in which the towns are authorized to contribute aid by way of taxation to any class of manufacturers, there is no difficulty in holding that this is not such a public purpose as we have been considering. If it be said that a benefit results to the local public of a town by establishing manufactures, the same may be said of any other business or pursuit which employs capital or labor. The merchant, the mechanic, the innkeeper, the banker, the builder, the steamboat owner are equally promoters of the public good, and equally deserving the aid of the citizens by forced contributions. No line can be drawn in favor of the manufacturer which would not open the coffers of the public treasury to the importunities of two-thirds of the business men of the city or town.

A reference to one or two cases adjudicated by courts of the highest character will be sufficient, if any authority were needed, to sustain us in this proposition.

In the case of *Allen v. The Inhabitants of Jay*, the town meeting had voted to loan their credit to the amount of $10,000, to Hutchins and Lane, if they would invest $12,000 in a steam saw-mill, grist-mill, and box-factory machinery, to be built in that town by them. There was a provision to secure the town by mortgage on the mill, and the selectmen were authorized to issue town bonds for the amount of the aid so voted. Ten of the taxable inhabitants of the town filed a bill to enjoin the selectmen from issuing the bonds.

The Supreme Judicial Court of Maine, in an able opinion by Chief Justice Appleton, held that this was not a public purpose, and that the town could levy no taxes on the inhabitants in aid of the enterprise, and could, therefore, issue no bonds, though a special act of the legislature had ratified the vote of the town, and they granted the injunction as prayed for.

Shortly after the disastrous fire in Boston, in 1872, which laid an important part of that city in ashes, the governor of the State convened the legislative body of Massachusetts, called the General Court, for the express purpose of affording some relief to the city and its people from the sufferings consequent on this great calamity. A statute was passed, among others, which authorized the city to issue its bonds to an amount not exceeding twenty millions of dollars, which bonds were to be loaned, under proper guards for securing the city from loss, to the owners of the ground whose buildings had been destroyed by fire, to aid them in rebuilding.

In the case of *Lowell v. The City of Boston*, in the Supreme Judicial Court of Massachusetts, the validity of this act was considered. We have been furnished a copy of the opinion, though it is not yet reported in the regular series of that court. The *American Law Review* for July, 1873, says that the question was elaborately and ably argued. The court, in an able and exhaustive opinion, decided that the law was unconstitutional, as giving a right to tax for other than a public purpose.

The same court had previously decided, in the case of *Jenkins v. Anderson*, that a statute authorizing the town authorities to aid by taxation a school established by the will of a citizen, and governed by trustees selected by the will, was void because the school was not under the control of the town officers, and was not, therefore, a public purpose for which taxes could be levied on the inhabitants.

The same principle precisely was decided by the State court of Wisconsin in the case of *Curtis v. Whipple*. In that case a special statute which authorized the town to aid the Jefferson Liberal Institute was declared void because, though a school of learning, it was a private enterprise not under the control of the town authorities. In the subsequent case of *Whiting v. Fond du Lac*, already cited, the principle is fully considered and reaffirmed.

These cases are clearly in point, and they assert a principle which meets our cordial approval.

We do not attach any importance to the fact that the town authorities paid one instalment of interest on these bonds. Such a payment works no estoppel. If the legislature was without power to authorize the issue of these bonds, and its statute attempting to confer such authority is void, the mere payment of interest, which was equally unauthorized, cannot create of itself a power to levy taxes, resting on no other foundation than the fact that they have once been illegally levied for that purpose.

The act of March 2d, 1872, concerning internal improvements, can give no assistance to these bonds. If we could hold that the corporation for manufacturing wrought-iron bridges was within the meaning of the statute, which seems very difficult to do, it would still be liable to the objection that money raised to assist the company was not for a public purpose, as we have already demonstrated.

JUDGMENT AFFIRMED.

Mr. Justice CLIFFORD, dissenting:

Unable to concur either in the opinion or judgment in this case, I will proceed to state, in very brief terms, the reasons which compel me to withhold my concurrence.

Corporations of a municipal character are created by the legislature, and the legislature, as the trustee or guardian of the public interest, has the exclusive and unrestrain control over such a franchise, and may enlarge, diminish, alter, change, or abolish the same at pleasure. Where the grantees of a franchise, as well as the grantors, are public bodies and the franchise is created solely for municipal objects, the grant is at all times within the control of the legislature, and consequently the charter is subject to amendment or repeal at the will of the granting power. Errors of indiscretion which the legislature may commit in the exercise of the power it possesses cannot be corrected by the courts, for the reason that the courts cannot adjudge an act of the legislature void unless it is in violation of the Federal or State constitution.

State constitutions may undoubtedly restrict the power of the legislature to pass laws, and it is plain that any law passed in violation of such a prohibition is void, but the better opinion is that where the constitution of the State contains no prohibition upon the subject, express or implied, neither the State nor Federal courts can declare a statute of the State void as unwise, unjust, or inexpedient, nor for any other cause, unless it be repugnant to the Federal Constitution. Except where the Constitution has imposed limits upon the legislative power the rule of law appears to be that the power of legislation must be considered as practically absolute, whether the law operates according to natural justice or not in any particular case, for the reason that courts are not the guardians of the rights of the people of the State, save where those rights are secured by some constitutional provision which comes within judicial cognizance; or, in the language of Marshall, C.J., "The interest, wisdom, and justice of the representative body furnish the only security in a large class of cases not regulated by any constitutional provision."

Courts cannot nullify an act of the State legislature on the vague ground that they think it opposed to a general latent spirit supposed to pervade or underlie the constitution, where neither the terms nor the implications of the instrument disclose any such restriction. Such a power is denied to the courts, because to concede it would be to make the courts sovereign over both the constitution and the people, and convert the government into a judicial despotism.

Subject to the Federal Constitution the legislature of the State possesses the whole legislative power of the people, except so far as the power is limited by the State constitution.

Our own decisions are to the same effect, as appears by one of very recent date, in which the court say that "the legislative power of a State extends to everything within the sphere of such power, except as it is restricted by the Federal Constitution or that of the State." ...

Unwise laws and such as are highly inexpedient and unjust are frequently passed by legislative bodies, but there is no power vested in a Circuit Court nor in this court, to determine that any law passed by a State legislature is void if it is not repugnant to their own constitution nor the Constitution of the United States.

Vague apprehensions seem to be entertained that unless such a power is claimed and exercised inequitable consequences may result from unnecessary taxation, but in my judgment there is much more to be dreaded from judicial decisions which may have the effect to sanction the fraudulent repudiation of honest debts, than from any statutes passed by the State to enable municipal corporations to meet and discharge their just pecuniary obligations.

[T]he obligation of the state to help, is limited to those who are unable to help themselves

---

STATE V. OSAWKEE TOWNSHIP

14 Kan.418 (Supreme Court of Kansas) 1875

Introduction

Although most of the focus of this book is on federal legislation, the U.S. retains a federal form of government in which state and local authorities often taken the lead in local disasters. This case from Kansas demonstrates that justices, in this case David Brewer (1837-1910), who would later serve on the U.S. Supreme Court, often interpreted state constitutions quite restrictively, in this case finding that government bonds to supply seeds to farmers was an unconstitutional form of class legislation.

BREWER, J.: But a single question is presented in this case for our consideration, and that is, the constitutionality of the act of the last legislature entitled "An act authorizing townships to issue bonds for relief purposes." (Laws of 1875, p. 53, ch. 42.) The matter has been pressed upon our early attention and decision for these reasons: The time within which these bonds may be issued is limited, the purposes sought to be accomplished thereby must be speedily accomplished. An impression widely prevails, supported by an official opinion of the attorney general, that the act is beyond the scope of the legislative authority, and that the bonds provided for in said act would, if issued, be destitute of legal obligation. Hence it is said, and with great propriety, that an authoritative decision is of public importance; that if the act be constitutional, such townships as desire may avail themselves of its benefits, and negotiate more easily and at higher figures the bonds they may issue, and that, on the other hand, if the act be unconstitutional no steps may be taken under it, the evil of repudiation be avoided, and other measures of relief be resorted to. Impressed with the force of these considerations, we have given the matter our early attention, and proceed now to state briefly the conclusions we have reached.

Two propositions may be considered settled: first, that taxation to be sustained must be for a public purpose; and second, that where municipal bonds are issued, whose payment is provided for solely by taxation, their validity depends upon the question whether the purposes to which the proceeds of such bonds are to be applied are public purposes. *Leavenworth County v. Miller*, 7 Kan. 479; *The Citizens Savings and Loan Association v. The City of Topeka*, recently decided by the supreme court of the United States; (23 Wallace.) It is also conceded by counsel that the entire purpose, or, if there are several, and no rule of apportionment as to the application of the proceeds, that all the purposes must be public. In other words, that the legislature cannot validate bonds for private purposes by declaring that the authorities may apply an indefinite portion of the proceeds to some public purpose. With these preliminary remarks let us turn to the act in question, and see to what purposes the proceeds of the bonds authorized by it are to be applied. The first four sections provide for the amount of bonds that may be issued, their form, title, time, rate of interest, the limit of the price for which they may be sold, and the placing of the proceeds to the credit of the relief fund. The last clause of section four then reads: "*Provided*, that no part of such fund shall be used except for the specific objects hereinafter named." Sec. 5 is as follows:

SEC. 5. The trustee, clerk and treasurer of such township, or a majority of them, shall, as soon as practicable, sell and dispose of the bonds issued by them under the authority of this act to the best possible advantage, and invest the proceeds, or so much thereof as in the judgment of said officers may be necessary, for the purpose of providing the destitute citizens of such townships with provisions and with grain for seed and feed; and the officers aforesaid shall distribute such articles of necessity amongst the destitute citizens of such township in proportion to their several necessities, under such rules and regulations as may be prescribed, in accordance with the provisions of the fourth section of this act: *Provided*, That no family shall receive more than seventy-five dollars in value."

The relief of the poor, the care of those who are unable to care for themselves, is among the unquestioned objects of public duty. In obedience to the impulses of common humanity, it is everywhere so recognized. Our own constitution but gives utterance to the universal voice when it says, "The respective counties of the state shall provide, as may be prescribed by law, for those inhabitants who, by reason of age, infirmity, or other misfortune may have claims upon sympathy and aid of society." Art. 7, § 4. It must be borne in mind however that the term "poor" is used in two senses. We use it in one sense simply as opposed to the term "rich." Thus we speak of the ordinary laborers, mechanics and artisans as poor people, without a thought of describing persons who are other than self-supporting. Indeed, the large majority of our people are poor people, and yet they would feel insulted to be told that they are objects of public charity. We use the term also to describe that class who are entirely destitute and helpless, and therefore dependent upon public charity. The dictionaries recognize this two-fold sense. Thus, Webster gives these definitions: "1. Destitute of property; wanting in material, riches, or goods; needy, indigent. It is often synonymous with indigent, and with necessitous, denoting extreme want. It is also applied to person who are not entirely destitute of property, but who are not rich; as, a poor man or woman; poor people. 2. (*Law.*) So completely destitute of property as to be entitled to maintenance from the public." Now, when we speak of the relief of the poor as a public duty, and one which may justify taxation, we use the term only in the latter sense. We have no thought of asserting that because a man is not rich, or even because he has nothing but the proceeds of his daily labor, therefore taxation may be upheld in his behalf. Such taxation would be simply an attempt on the part of the state to equalize the property of its citizens. Something more than poverty, in that sense of the term, is essential to charge the state with the duty of support. It is, strictly speaking, the pauper, and not the poor man, who has claims on public charity. It is not one who is in want merely, but one who, being in want, is unable to prevent or remove such want. There is the idea of helplessness as well as of destitution. We speak of those whom society must aid, as the dependent classes, not simply because they do depend on society, but because they cannot do otherwise than thus depend. Cold and harsh as the statement may seem, it is nevertheless true, that the obligation of the state to help, is limited to those who are *unable* to help themselves. It matters not through what the inability arises, whether from age, physical infirmity, or other misfortune; it is enough that it exists. It is doubtless true, that in the actual administration of the poor-laws, many who are not properly entitled thereto receive public support; but failures in the administration of laws do not change the principles upon which they must rest. It is important to bear this distinction in mind, for, as will appear hereafter, it is really the former and not the latter class which is sought to be relieved under this law. It may be remarked in passing, that it was claimed by counsel as one of the objections to this act, that under the rule, "*expressio unius, exclusio alterius*," inasmuch as the constitution casts upon the respective counties the care of the destitute, there was an implied prohibition upon casting it elsewhere. Much might be said, and with great force, in support of this objection; but we do not care to decide whether it be well taken or not, much less to rest this case upon it, for such a decision might be construed as an implied recognition of the validity of the principle which we are constrained to believe cannot be sustained.

The purpose of the act, as expressed in the section quoted, is to provide the destitute with provisions and with grain for seed and feed. This legislation must be construed in the light of known facts. For reasons unnecessary here to recount, in some portions of the state last season there was a total, and in others a partial failure of the crops. It was generally understood that many farmers would come to this spring's sowing with little or no seed, and with stock weakened for lack of grain. To make good this lack is the evident purpose of the act, to provide grain for seed and feed. Its aim is not to furnish food to the hungry, clothing to the naked, or fuel to those suffering from cold. It is not the helpless and dependent, whose wants are alone sought to be relieved. If it were, the fact that many who are neither helpless nor dependent might obtain assistance through its administration, would be no valid objection to the constitutionality of the law. It contemplates a class who have fields to till and stock to care for, and purposes to help them with seed for their fields and grain for their stock, that thus they may pursue with better prospects of success their ordinary avocations. It taxes the whole community to assist one class, and that not for the purpose of relieving actual want, but to assist them in their regular occupations. These people are engaged in the business of farming. This business cannot be successfully carried on without seed, nor without stock strong enough to do the ordinary work. They are destitute of seed, and their stock require grain. Hence the tax upon the community. The principle would be the same if their supply of grain was sufficient, but through the prevalence of the epizooty, or some other disease, their stock had all died. Could a tax be sustained to purchase stock for their ordinary farm work? Or again, suppose some prairie fire driven by a fearful wind sweeps through a county, consuming its fences and farming tools: can a tax be sustained to supply this loss, and enable the farmers to prosecute their labors? Nor need the inquiry be limited to a single class. Were the carpenters or shoemakers, or any other industrial class, located in a separate quarter of a city, and their tools and stock in trade swept away by fire, could a tax be sustained to purchase new sets of tools and new stock in trade to enable them to reprosecute their business and secure support for themselves and families? No distinction in principle can be made between these different supposed cases, and the case at bar. They all rest upon this proposition, that a tax is laid upon the public to furnish to one class the means of carrying on its regular occupation. A further examination of this act will but strengthen the views herein expressed. The four succeeding sections are as follows: ... [Brewer quotes various sections of the law.]

These various provisions show that the idea of the legislature was not the relief of the helpless and dependent, but the assistance of a class temporarily embarrassed. The recipient is required to make oath that he is buying the aid for himself, and not on a speculation. He is to give a note for the amount received, and if a married man the note must also be signed by his wife. The note is to bear the same date and draw the same interest as the bonds, and the interest is payable at the same time as the interest on them. This note is to be a mortgage as well, and the most sweeping kind of a mortgage too, embracing all the real and personal property of the maker, whether owned at the time of its execution or subsequently acquired. And finally, it is made the express duty of the township treasurer to see to the collection of this note, and to take all proper and needful action therefor. Nothing is contemplated but a loan, and a secured loan at that. The credit of the township is invoked to procure funds for the accommodation of a single class temporarily and through unexpected calamity embarrassed in the prosecution of its ordinary business. Can this be called a public purpose? Clearly not. It would doubtless relieve the temporary wants of that class, would enable it to enter upon the business of the year with increased hope and a reasonable expectation of ordinary success in that business, and thus indirectly result in great benefit to the general public. But a similar result would follow the success and prosperity of any other class in business. And if the principle be once recognized in its application to this class, who can tell how soon it may be invoked in aid of another? If one hundred farmers may receive seventy-five dollars each to assist them in their farming, why may not one hundred mechanics with equal propriety receive seventy dollars each to assist them in their business? or a single manufacturer who employs one hundred hands receive seventy-five hundred dollars to assist him in his manufacturing? A difference in amount makes no difference in the principle.

But it may be said that this legislation can be defended as preventive and anticipatory. To prevent the spread of disease quarantine regulations are enforced, and ships coming from certain places are with all their passengers detained in quarantine even when not a solitary case of sickness exists on board. To prevent ignorance in the voter, the child is compelled to be educated. To prevent crime in the man, the boy is sent to the reform school. To prevent the spread of fire, valuable buildings are pulled down and destroyed. Grant that these parties are not now helpless and dependent; that they are not a public charge. Unless they are able to make and harvest a crop they may become so the ensuing winter. Is it not the part of wisdom to expend a little now to purchase seed and feed, rather than run the risk of having them become paupers hereafter? Under the peculiar circumstances of this case, this argument is a strong one. We are not disposed to belittle the magnitude of the calamity, or make light of the hardships of those upon whom it has principally fallen. If we consulted simply our own feelings we should gladly approve of this, as of every effort to mitigate the severity of the blow. But, though this calamity is great, and though by reason thereof it may seem wise to appropriate out of the public funds a little now to guard against the risk of future want, yet the principle is dangerous and unsound. Let the doorways of taxation be opened, not merely to the relief of present and actual distress, but in anticipation of and to guard against future want, and who can declare the result? How certain must be the expectation of want? how nigh its approach? What efforts must the individual make to ward it off? May he do nothing, and demand that the public make provision to guard against the possibility of future suffering? Must widespread and general calamity precede the granting of such anticipatory relief, or is it enough that individual misfortune or indolence render probable the approach of want? The mere mention of these questions suggests the dangers which would follow the adoption of this as a rule of public conduct. But the attendant dangers of such a rule are not the sole or the controlling considerations. The relief provided in this act is only indirect, and contingent. There is no direct appropriation to meet future want. The appropriation is for present use, and the relief is contingent on the successful prosecution of the business of the recipients during the ensuing year. If the crop proves a failure, the public funds are lost, and no relief is secured. It is a speculation, which however proper and reasonable for individuals, is not a legitimate part of public duty. The same principle would justify assistance to a mechanic destitute of tools, to enable him to purchase tools, and through their use in his regular calling prevent his becoming a public burden. Indeed, it would be difficult to deny its application in any case where by present assistance, either in the purchase of implements or stock in trade, the recipient might reasonably be expected to earn a subsistence in the prosecution of his regular business. We should expect to find but few authorities to throw any light upon this question. The case of the *Citizens Savings and Loan Association v. The City of Topeka,* heretofore cited, decides that taxation cannot be invoked to assist private manufacturing establishments. The propositions laid down by Mr. Justice Miller in reference thereto are broad. It matters not how great may be the necessities of such an establishment, or how much it may indirectly benefit the community, it cannot be aided by taxation. The same propositions were asserted by the supreme court of Maine in the case of *Allen v. Inhabitants of Jay*, 60 Me. 124. That was a case of an attempted loan of the credit of a town to certain parties in consideration of their engaging in some manufacturing enterprises for their private emolument. In delivering the opinion of the court C. J. Appleton uses this strong language: "But whether the money raised is to be distributed *per capita*, or loaned, can make no difference in principle. If towns can assess and collect money to be again loaned to such persons as the majority may select for such purposes as it may favor, with such security, or without security, as it may elect, property ceases to be protected in its acquisition or enjoyment. * * * If the loan be made to 429 one or more for a particular object, it is favoritism. It is a discrimination in favor of the particular individual, and a particular industry thereby aided, and is one adverse to and against all individuals, all industries, not thus aided. If it is to be loaned to all, then it is practically a division of property under the name of a loan. It is communism incipient, if not perfected." But the case most nearly in point is that of *Lowell v. The City of Boston*, recently decided by the supreme court of Massachusetts. In that as in this, it was the circumstance of a great public calamity, the Boston fire, and a praiseworthy effort on the part of the legislature to provide assistance for the sufferers thereby. An act was passed authorizing the city of Boston to loan its credit to assist in rebuilding the burnt district. But this was declared to be outside the purposes for which taxes could be levied, or bonds issued. We have been able to find nothing more in point than these authorities, and they all point in the same direction as the considerations we have heretofore adverted to.

It is with reluctance that we have reached the conclusion that this act cannot be sustained. But ours is an unmixed duty, to declare the law as it is, and not as we might wish it to be. Especially imperative is that duty when as in cases like the present there is in the surrounding circumstances a strong appeal to overlook permanent rules in favor of a present and pressing want.

The judgment of the court below is reversed.

All the Justices concurring.

Whoever ... uses any part of the Army ...
as a posse comitatus ... shall be fined

POSSE COMITATUS ACT OF 1878

Introduction

Prior to the institutionalization of police forces, it was common to call upon citizens to hunt down fugitives or restore order. This was known as a posse comitatus, meaning "power of the county." In early U.S. history, military forces were sometimes also used for such purposes.

After the South's bitter experience with federal military rule during the period of Congressional Reconstruction (1865-1877), which followed the Civil War, Congress adopted the Posse Comitatus Act of 1878. It recognized the intricacies of the federal system by typically preventing the use of federal troops to enforce domestic laws, although some exceptions have been made in order to use military resources to combat illegal drug trafficking. This limitation typically applies during natural disasters although it does not apply to National Guard units, as long as they remain under the control of the state governor. This generally works against attempts to federalize the Guard during such emergencies.

Section 18. (a) title 18, United States Code is amended by inserting the following new section after section 1384:

"Sec. 1835. Use of Army and Air Force as posse comitatus

"Whoever, except in cases and under circumstances expressly authorized by the Constitution or Act of Congress, willfully uses any part of the Army or the Air Force as a posse comitatus or otherwise to execute the laws shall be fined not more than $10,000 or imprisoned not more than two years or both. This section does not apply in Alaska."

PART IV

From the End of the Nineteenth Century through World War I

Men, therefore, owe to men, in the chances and perils of this life, aid and sympathy, on account of the common participation in human frailty and folly

---

William Graham Sumner
WHAT SOCIAL CLASSES OWE EACH OTHER

1883

Introduction

William Graham Sumner (1849-1910), who began his career as an Episcopalian priest, became a professor of political and social science at Yale and earned a reputation as one of the founders of sociology. Sumner was a disciple of Herbert Spencer and a proponent of Social Darwinism, who saw class divisions as inevitable and the role of government as minimal (his philosophy is close to that of modern libertarians). In this selection from the last chapter of one his most famous books, Spencer distinguishes the role of private charity from that of government. The prevalence of this philosophy is one explanation for America's early reliance on private charity for disaster relief rather than governmental aid.

From Chapter XI, Wherefore We Should Love One Another

In the preceding chapters I have discussed the public and social relations of classes, and those social topics in which groups of persons are considered as groups or classes, without regard to personal merits or demerits. I have relegated all charitable work to the domain of private relations, where personal acquaintance and personal estimates may furnish the proper limitations and guarantees. A man who had no sympathies and no sentiments would be a very poor creature; but the public charities, more especially the legislative charities, nourish no man's sympathies and sentiments. Furthermore, it ought to be distinctly perceived that any charitable and benevolent effort which any man desires to make voluntarily, to see if he can do any good, lies entirely beyond the field of discussion. It would be as impertinent to prevent his effort as it is to force cooperation in an effort on some one who does not want to participate in it. What I choose to do by way of exercising my own sympathies under my own reason and conscience is one thing; what another man forces me to do of a sympathetic character, because his reason and conscience approve of it, is quite another thing.

What, now, is the reason why we should help each other? This carries us back to the other illustration with which we started [that of a man encountering someone pinned by a fallen tree and lecturing him on the laws of gravity]. We may philosophize as coolly and correctly as we choose about our duties and about the laws of right living; no one of us lives up to what he knows. The man struck by the falling tree has, perhaps, been careless. We are all careless. Environed as we are by risks and perils, which befall us as misfortunes, no man of us is in a position to say, "I know all the laws, and am sure to obey them all; therefore I shall never need aid and sympathy." At the very best, one of us fails in one way and another in another, if we do not fail altogether. Therefore the man under the tree is the one of us who for the moment is smitten. It may be you tomorrow, and I next day. It is the common frailty in the midst of a common peril which gives us a kind of solidarity of interest to rescue the one for whom the chances of life have turned out badly just now. Probably the victim is to blame. He almost always is so. A lecture to that effect in the crisis of his peril would be out of place, because it would not fit the need of the moment; but it would be very much in place at another time, when the need was to avert the repletion of such an accident to somebody else. Men, therefore, owe to men, in the chances and perils of this life, aid and sympathy, on account of the common participation in human frailty and folly. This observation, however, puts aid and sympathy in the field of private and personal relations, under the regulation of reason and conscience, and gives no ground for mechanical and impersonal schemes. [Sumner proceeds to make four distinctions]. ...

4. The law of sympathy, by which we share each others' burdens, is to do as we would be done by. It is not a scientific principle, and does not admit of such generalization or

interpretation that A can tell B what this law enjoins on B to do. Hence the relations of sympathy and sentiment are essentially limited to two persons only, and they cannot be made a basis for the relations of groups of persons, or for discussion by any third party.

Social improvement is not to be won by direct effort. It is secondary, and results from physical or economic improvements. That is the reason why schemes of direct social amelioration always have an arbitrary, sentimental, and artificial character, while true social advance must be a product and a growth. The efforts which are being put forth for every kind of progress in the arts and sciences are, therefore, contributing to true social progress....

The only help which is generally expedient, even within the limits of the private and personal relations of two persons to each other, is that which consists in helping a man to help himself. This always consists in opening the chances. A man of assured position can by an effort which is of no appreciable importance to him, give aid which is of incalculable value to a man who is all ready to make his own career if he can only get a chance. The trues and deepest pathos in this world is not that of suffering but that of brave struggling. The truest sympathy is not compassion, but a fellow-feeling with courage and fortitude in the midst of noble effort.

Now, the aid which helps a man to help himself is not in the least akin to the aid which is given in charity. If alms are given, or if we "make work" for a man, or "give him employment," or "protect" him, we simply take a product from one and give it to another. If we help a man to help himself, by opening the chances around him, we put him in a position to add to the wealth of the community by putting new powers in operation to produce. It would seem that the difference between getting something already in existence from the one who has it, and producing a new thing by applying new labor to natural materials, would be so plain as never to be forgotten; but the fallacy of confusing the two is one of the commonest in all social discussions.

Source: https://mises/org/library/what-social/=-classes-owe-each-other -0. 12/13/2016.

Necessity knows neither law nor constitution, and never did in this country

---

DEBATE OVER AID FOR FLOOD VICTIMS ALONG OHIO RIVER

February 11, 1884

Introduction

The following excerpts are from debates over whether to extend federal aid to individuals who had been made homeless by a flood on the Ohio River and its tributaries in 1884. Congressmen continued to disagree about the constitutionality of aid, especially to states that they thought were capable of taking care of their own. Many relied on earlier relief precedents, which they cited. Although most proponents of aid sought to justify the constitutionality of relief agencies, many argued that necessity might trump legal justifications during dire circumstances.

Resolved, &c., That the sum of $300,000 is hereby appropriated, or so much thereof as may be necessary out of any money in the Treasury not otherwise appropriated, to be expended by and under the direction of the Secretary of War in the purchase and distribution of subsistence stores, clothing and other necessary articles to such destitute persons as may require assistance in the district overflowed by the Ohio River and its tributaries by the recent flood; and he is authorized to co-operate with the several States and cities of which said district is a part in making distribution of the same. And the Secretary of War is also authorized in his discretion to use the hospital and other tents now possessed by the War Department so far as may be necessary to furnish temporary shelter to the destitute people who have been driven from their homes by said overflow, and is authorized to use steamboats and other boats and vessels belonging to and now employed by the Government upon the Ohio and Mississippi Rivers and their tributaries in the transportation and distribution of the supplies furnished by the United States or individuals to and among said destitute and suffering people; and he may employ such other means of transportation as he may deem necessary to carry the purposes of this resolution into effect. (1032)

Mr. [John Fassett] FOLLETT [OH]. I will state very briefly, Mr. Speaker, the reasons for this appropriation. The gentleman from New York [Mr. HISCOCK] desires to know whether there is any warrant in the Constitution for such an appropriation. Necessity knows neither law nor constitution, and never did in this country. [Laughter and applause on the Republican side.] There are precedents of this House and of Congress, for exactly such appropriations. The precedent was set in the last Congress by an appropriation for the relief of the sufferers along the Mississippi River. There has never been in the history of this country such a flood as that which is now raging upon the Ohio River. Families by the thousands are driven from their homes, and are to-day camping out in open fields with nothing to protect them except simply such clothing as they could save from the wrecks of their houses. The cities and towns along the line of the Ohio River and most of its tributaries are now under water. I have just received a dispatch from the city of Cincinnati, which I desire to read, coming as it does, from the president of one of our banks and one of the leading business men of the city:

Do all you can to get appropriation for flood sufferers. Water still rising fast, and there will be great loss and distress.

Thousands of families along the Ohio River are now houseless and homeless. Thousands of families are dependent entirely upon charity—charity of neighbors situated in great measure exactly as they are, driven from their homes, and having no means to furnish subsistence and support to those sufferers.

Mr. [Frank] HISCOCK [NY]I would like to ask the gentleman from Ohio whether any application has been made by the Legislature of Ohio for this relief from Congress, and whether the State, through its Legislature, has made any appropriation for the purpose?

Mr. FOLLETT. The State of Ohio, as I am informed, has made an appropriation. The Chamber of Commerce of Cincinnati has made an appropriation; and I can say to the gentleman that so far as Cincinnati is concerned, I think she will take care of her people. But Cincinnati is not the only place that is overwhelmed by this misfortune. A town but fifteen miles above that city is entirely under water; there is not a house from which subsistence can be furnished to the sufferers. There is no means by which it can be afforded, except by outside aid. A town just below Cincinnati—the town of Lawrenceburg—is also entirely under water, and there is nobody there so situated as to render relief.

I can say to the gentleman and to this House that the State which I have the honor in part to represent here never turned a deaf ear to the voice of suffering. When the cry came one year ago from across the waters—from Germany—that her people were suffering in consequence of floods, our people sent them assistance, not by thousands simply but by the fifty thousands of dollars.

Congress is now asked to furnish temporary relief to these people who have become involved in a calamity unprecedented since this country was first settled. It is true that one year ago we had in this country a calamity of a similar character, but not of like extent with this. At that time the State of Ohio made its contributions for the aid of the sufferers. The Chamber of Commerce of Cincinnati and every moneyed institution of the place not only contributed, but organized their charity in such a way that it was distributed so as most effectually to relieve the suffering. And to-day the only subsistence these sufferers have is such as is furnished by these charitable and good-hearted people.

Mr. Speaker, under such circumstances this is not a question of what the letter of the Constitution or the letter of the law may be. The cry of distress comes up for relief which can be furnished, and it is the duty of Congress when such relief is in its power to furnish it.

Mr. Speaker, the people involved in this calamity number hundreds of thousands. It is spreading along the Ohio River and will reach its mouth. Every tributary of that river has had a flood of water unknown in history. At Wheeling, where it is now falling, it was nine feet higher than ever before known since the country was settled.

These people, as I said before, are now houseless ahd homeless. You ask they shall be relieved by their immediate neighbors. Those neighbors are granting relief to the greatest extent in their power. But shall the great Government of the United States, with its Treasury overflowing with a surplus there that can be appropriated, sit idly by when our citizens are suffering, our people are starving, and to-day without shelter from the storms still pitilessly beating upon their heads? [Applause.] (1033)

Mr. HISCOCK. I Remember the State of Illinois came knocking at the door of the Treasury for assistance, and if there ever was an occasion when I admired my friend from Illinois [Mr. CANNON], who was with me on the Committee on Appropriations, it was when he shrugged his shoulders at his State for making that application. Where States are poor, where there is a large indigent population, we will say, like the negroes, who are the ward of the nation, you may come here and ask to have them succored and supporter; but for Ohio, New York, Illinois, Michigan even, to pass around the hat of beggary, sir, I do not admire the spirit in either of those States which would permit it. And so far as my State is concerned, they would rally in their cities, rally in their churches—I believe they would prefer to do it for Ohio even than to have a Representative of that State stand here on this floor and advocate it to be taken out of the Federal Treasury. (1033)

Mr. CANNON [attempting to clarify the position that had been imputed to him].

We have been familiar with similar questions during the last Congress, and I think also during some Congresses prior to that. Appropriations similar to this have been made. The gentleman from New York, at that time the chairman of the Committee on Appropriations and my colleague on that committee, does state correctly that when the proposition was made to expend a portion of the money of the Unites States for the relief of the State of Illinois I was anxious that expenditure should be guarded. I did not want citizens of Illinois, nor do I want them now, to be relieved except as they may have needed immediate relief because of the necessities that were pressing upon them on account of a great emergency. The Legislature of the State was at that time in session, and the great

State of Illinois was equal to the task of relieving her citizens, as she appropriated $250,000 from her treasury ...But at that time some citizens along the Ohio and the Mississippi Rivers did receive temporary relief under the action of Congress by the issuing of rations, and perhaps in some instances clothing, before the relief could be given by the State.

Now, I apprehend, without being troubled by the constitution question so far as I am concerned, that when the waters of this country descend almost for forty days and forty nights, and men and women and children are driven by multiplied thousands from their homes and are without shelter and starving, I am ready to give them relief by appropriations from the Federal Treasury temporarily or until their States can act. And then I am frank to say that I want to see the great States, especially of the North, with their wealth and intelligence and liberality, come to the rescue and protect their citizens that are in want out of their own treasures and by their own means; and subject to this emergency I believe these States are ready and willing, including my own, to do that thing. (1034)

Mr. BAYNE. It seems to me there is no constitutional difficulty in the way of this appropriation. I think that the clause of the Constitution which authorizes the Congress of the United States "to lay and collect taxes, duties, imposts, and excises, to pay the debts and provide for the common defense and general welfare of the United States" includes the right to make this appropriation. If any great calamity shall befall any section of this country such as has befallen the people along the Ohio River and its tributaries, I have not a doubt in my mind about the constitutional power of Congress to make an appropriation from those taxes to meet that emergency and to provide for the welfare of those people.

On the same principle I am willing to justify the bill which was reported from the Committee on Agriculture to prevent the spread of pleuro-pneumonia and other contagious diseases among cattle. It is to provide for the general welfare. And that the power is in the Congress of the United States to make this appropriation, under a proper interpretation of the Constitution, I have not a particle of doubt.

Mr. Speaker, the district which I represent has been very largely a sufferer by this terrible calamity. There have been thousands and thousands of people driven from their homes by this great flood. A great many of them are now suffering and in want. Our people have raised voluntary contributions and our cities have made appropriations for the relief of these people. I am not prepared to say whether our citizens would take advantage of this appropriation; I am inclined to think our people would not; I believe they will adequately provide for the suffering that is being there endured by them.

But while I think that, I believe it is the bounden duty of every gentleman here to support this measure and to help these communities and localities which are unable to help themselves. There are certain places along the Ohio River which I know where the people are driven from their homes, and where it would be a burden that could not be endured by their neighbors and friends to raise the requisite amount of money or the requisite provisions for their comfort and safety during this time of suffering. Therefore I say we should make this appropriations. I say it should receive the hearty and unanimous concurrence of Congress, and that relief should be sent as soon as possible to these suffering people. And as to the constitutional difficulties, even if there were such, they should not stand in my way in making an appropriation. But, as I have said before, I believe there is no constitutional impediment in the way of this appropriation at all, and that we are perfectly justified in making it under the letter and in the spirit of the Constitution. (1034)

Mr. HORR. I want to say that the gentleman's colleague from New York who is now facing me {Mr. COX}, who has the most critical mind on constitutional questions in this House, will certainly agree with me that this bill comes within the pale of the Constitution, because he can not have forgotten that during, I think, the last session of Congress he carried around in his pocket a resolution to give to the starving Irish $300,000 under somewhat similar circumstances.

... after thinking of this matter a great deal I am firmly of the opinion that when there is any great calamity like the one that is now visiting a portion of our common country, it is within its constitutional power for the Government of the United States to step in and give temporary help....

My doctrine has always been that in construing the Constitution, where there is a great good to be done, where there is a great calamity to be averted, to construe it liberally; but where the rights of citizens are at stake I would construe it strictly. I believe that in that view of the matter hardly any man here can vote against our sending relief to that entire valley which to-day is under water and its citizens fleeing in every direction for shelter and food....

Mr. GOFF. ... In the face of this great calamity it is no time to split hairs on constitutional questions. If we have not the constitutional right to grant this relief, which I do not concede, then it may be based upon charity, as appealing to the finer instincts of humanity, in which the people of this country have never yet been found wanting. Basing it upon that, I say there should not be a dissenting vote in this House to-day on this question. It has grown to be a practice, it has become almost a custom. In the face of these great national calamities, for such appropriations as this to be made.

Let me say right here that on account of this fact private subscriptions and private donations have not poured in and will not pour in as they have in the past. (1035)

Mr. BELFORD. ...Talk about Constitution! Talk about laws! Humanity is greater than any constitution which was ever formulated by any people. It is humanity for which constitutions are enacted and laws are made. And Paul put it right when ye said that the greatest of all the virtues is charity, either in the individual or in the nation.

Talk about State rights in a matter in the cloudy presence of a constantly accumulating calamity! Suppose my friend from Connecticut [Mr. EATON] has lived in the days of Noah [laughter] and had found himself perched on the top of Mount Ararat looking down upon the submerged plains [renewed laughter]; the people fleeing from Nineveh, some swimming, others drowning , and the children lifting up their hands to him to render assistane; I suppose he would have referred that drowning crowd to the city council of Nineveh and told them to make an appropriation to stop the rain. [Laughter and applause.] That would have been State-rights doctrine then, and it is State-rights doctrine to-day.

I hold this is a nation, and a nation which is deeply interested in the welfare, happiness, and prosperity of every citizen who lives within its limits. ... (1036)

Mr. REAGAN. ...I only wish to add there is nothing in the theory of the Government to forbid what we are doing now; and while there may not be an express declaration of the Constitution that we may appropriate to relieve the calamities of the people involved in distress, we have a long line of precedents which have met the approbation of the most illustrious minds of the past, and we know that if we shall do what we are now asked to do we do not violate the Constitution or inflict wrong upon States or wrong upon the people; but simply representing the great Government of the Union, we say that out of its Treasury we propose to contribute something to the relief of sufferers who may be dying now for the want of it. (1037)

Mr. BLOUNT ... I have not forgotten that sympathy which was shown to my own section years ago. I have not forgotten the suffering of the people of my native State and how generously and unanimously the Congress of the United States responded to their necessity and suffering. Ay, more, I have not forgotten gotten when the yellow-fever attacked various cities along the Gulf of Mexico and the Mississippi River, and an appeal was made here for aid. I have not forgotten that when a proposition was before gthis House to appropriate—not by estimates, not by detailed statements as to the necessities of the situation—a half million of dollars to be used in the discretion of the Government, and to be used not merely during a period of three years, for the relief of those people of my section.... (1038)

Note: The House voted 234 to 23 (with 74 not voting) in favor of the measure. (See 1040)

[T]hough the people support the government the government should not support the people

Grover Cleveland
VETO OF TEXAS SEED BILL

February 16, 1887

Introduction

In the centennial year of the writing of the U.S. Constitution, President Grover Cleveland, one of the few Democratic presidents of this period, vetoed a special distribution of seeds to a drought-stricken area of Texas. Cleveland expressed the continuing concern over governmental paternalism, which he contrasted to the provision of private charity. He did suggest that states could forgo their own quotas of grain in order to help the stricken region.

To the House of Representatives:

I return without my approval House bill No. 10203, entitled "An act to enable the Commissioner of Agriculture to make a special distribution of seeds in the drought-stricken counties of Texas, and making an appropriation therefor."

It is represented that a long-continued and extensive drought has existed in certain portions of the State of Texas, resulting in a failure of crops and consequent distress and destitution.

Though there has been some difference in statements concerning the extent of the people's needs in the localities thus affected, there seems to be no doubt that there has existed a condition calling for relief; and I am willing to believe that, notwithstanding the aid already furnished, a donation of seed grain to the farmers located in this region, to enable them to put in new crops, would serve to avert a continuance or return of an unfortunate blight.

And yet I feel obliged to withhold my approval of the plan, as proposed by this bill, to indulge a benevolent and charitable sentiment through the appropriation of public funds for that purpose.

I can find no warrant for such an appropriation in the Constitution, and I do not believe that the power and duty of the General Government ought to be extended to the relief of individual suffering which is in no manner properly related to the public service or benefit. A prevalent tendency to disregard the limited mission of this power and duty should, I think, be steadfastly resisted, to the end that the lesson should be constantly enforced that though the people support the Government the Government should not support the people.

The friendliness and charity of our countrymen can always be relied upon to relieve their fellow-citizens in misfortune. This has been repeatedly and quite lately demonstrated. Federal aid in such cases encourages the expectation of paternal care on the part of the Government and weakens the sturdiness of our national character, while it prevents the indulgence among our people of that kindly sentiment and conduct which strengthens the bonds of a common brotherhood.

It is within my personal knowledge that individual aid has to some extent already been extended to the sufferers mentioned in this bill. The failure of the proposed appropriation of $10,000 additional to meet their remaining wants will not necessarily result in continued distress if the emergency is fully made known to the people of the country.

It is here suggested that the Commissioner of Agriculture is annually directed to expend a large sum of money for the purchase, propagation, and distribution of seeds and other things of this description, two-thirds of which are, upon the request of Senators, Representatives, and Delegates in Congress, supplied to them for distribution among their constituents.

The appropriation of the current year for this purpose is $100,000, and it will probably be no less in the appropriation for the ensuing year. I understand that a large quantity of grain is furnished for such distribution, and it is supposed that this free apportionment among their neighbors is a privilege which may be waived by our Senators and Representatives.

If sufficient of them should request the Commissioner of Agriculture to send their shares of the grain thus allowed them to the suffering farmers of Texas, they might be enabled to sow their crops, the constituents for whom in theory this grain is intended could well bear the temporary deprivation, and the donors would experience the satisfaction attending deeds of charity.

Source: Website of Miller Center at the University of Virginia.

Is it proper to make appropriations for ... unfortunate victims ...?

John Randolph Tucker
THE CONSTITUTION OF THE UNITED STATES
A CRITICAL DISCUSSION OF ITS GENESIS, DEVELOPMENT, AND INTERPRETATION
1899

Introduction

Questions about the legitimacy of governmental spending on social programs continued into the twentieth century. In contrast to the relatively liberal views that St. George Tucker had articulated in a House Committee report, his grandson John Randolph Tucker, a former president of the American Bar Association who taught at Washington and Lee University in Virginia, continued to articulate a more restrictive view of the General Welfare Clause. This was reflected in a book that he published in 1899. His son, Harry, would later published an article entitled "The General Welfare," in a 1922 issue of the *Virginia Law Review* in which he advocated a similar view, which critics often used to criticize New Deal programs.

§ 230. ...To speak of the government as a government of granted powers, and yet to hold that its greatest power, that of unlimited taxation, can be applied to the execution of powers not granted nor enumerated in the Constitution, but to such objects as Congress may deem to be for the common defense and general welfare, would make it substantially a government of unlimited powers. It is well said by Mr. Justice Miller: "Of all the powers conferred upon government, that of taxation is the most liable to abuse. Given a purpose or object for which taxation may be lawfully used, and the extent of its exercise is in its very nature unlimited."

The theory of construction which we are combating involves the historic conclusion that the States, which were jealously guarding their reserved powers by expressly enumerating the powers delegated to the Federal government, agreed under this power of taxation, in the face of its liability to abuse by exercise for objects and purposes which were ungranted, that Congress might nevertheless, at its discretion, assume and execute them through the power of appropriation. It is absurd to suppose that such was the purpose of the States which framed the Constitution. ...

§232. Let us now consider some of the forms in which this question has presented itself.

1st. Can bounties on products be paid, in the form of appropriations, to aid one or more of the interests of private persons? Does the fact that such industries aid partially the general welfare justify the public aid through taxation to mere personal industries? Is it proper to make appropriations for the unfortunate victims of overflows, fire, grasshoppers, and the like, in order to promote the general welfare? Does help by public money to individuals who may need it elevate such a case to the promotion of the general welfare? Do a number of such particular cases constitute general welfare of private individuals; but are not these appropriations for private benefit the principal object, while the public welfare is merely the incident?

...§ 233. Several cases are cited by the Justice in illustration of this. A town meeting voted the town's credit to the amount of ten thousand dollars to certain individuals if they would invest twelve thousand dollars in certain mills, etc., to be built in that town by them. Provision was made to secure the town by mortgage on the mills. The Supreme Court of Maine held that this was not a public purpose, and the town could levy no taxes in aid of the enterprise, and could issue no bonds for the purpose, though an act of the legislature had ratified the vote of the town. The disastrous fire in Boston in 1872 gave rise to the passage of a law by the legislature of Massachusetts which authorized the city to issue its bonds which were to be loaned under proper security by the city to owners of the ground on which the buildings had been destroyed by fire, to aid them in rebuilding. The court held the law unconstitutional as authorizing taxation for what was not a public purpose. [Citation to *Lowell v. Boston,* 111 Mass. 484.]

Source: Tucker, John Randolph, *The Constitution of the United States: A Critical Discussion of Its Genesis, Development, and Interpretation*. Ed., by Henry St. George Tucker, 2 vols. Chicago: Callaghan & Co., 1899), I, 490-491, 493, 496.

To act in matters of voluntary relief

THE AMERICAN NATIONAL RED CROSS CHARTER OF 1900

31 Stat. 277-280

Introduction

American Disaster relief is often a coordinated effort not between the legislative and executive branches, and the nation and the states, but also among the government, private individuals, and nongovernmental agencies, especially those motivated by religious and humanitarian impulses. Historically, groups responsible for aiding victims of natural disasters also seek to aid victims of war. The Red Cross actually grew out of the efforts of Clara Barton and others to care for wounded soldiers during the U.S. Civil War. This charter, which has been revised a number of times, shows this connection. The Red Cross remains the only nongovernmental organization that the U.S. government has chartered for such purposes and to which the government regularly gives aid. This sometimes blurs the line between its private and governmental roles.

American National Red Cross, Incorporated. Preamble.

Whereas on the twenty-second of August, eighteen hundred and sixty-four, at Geneva, Switzerland, plenipotentiaries respectively representing Italy, Baden Belgium, Denmark, Spain, Portugal, France, Prussia, Saxony, and Wurtemburg, and the Federal Council of Switzerland agreed upon ten articles of a treaty or convention for the purpose of mitigating the evils inseparable from war; of suppressing the needless severity and ameliorating the condition of soldiers wounded on the field of battle; and particularly providing, among other things, in effect, that persons employed in hospitals, and in affording relief to the sick and wounded, and supplies for this purpose, shall be deemed neutral and entitled to protection; and that a distinctive and uniform flag shall be adopted for hospitals and ambulances, and convoys of sick and wounded, and an arm badge for individuals neutralized; and

Whereas said treaty has been ratified by all of said nations, and by others subsequently, to the number of forty-three or more, including the United States of America; and

Whereas a permanent organization is an agency needed in every nation to carry out the purposes of said treaty, and especially to secure supplies and to execute the humane objects contemplated by said treaty, with the power to adopt and use the distinctive flag and arm badge specified by said treaty in article seven, on which shall be the sign of Red Cross, for the purposes of cooperating with the "Comite International de Secours aux Militaires Blesses" (International Committee of Relief for the Wounded in War); and

Whereas, in accordance with the requirements and customs of said international body, such an association, adopting and using said insignia, was formed in the city of Washington, District of Columbia, in July, eighteen hundred and eighty-one, known as "The American National Association of the Red Cross," and reincorporated April seventeenth, eighteen hundred and ninety-three, under the laws of the District of Columbia; and

Whereas it is believed that the importance of the work demands a reincorporation by the Congress of the United States: Now, therefore,

Incorporators.

Be it enacted by the Senate and House of Representatives of the United States of America in Congress assembled, That Clara Barton, George Kennan, Julian B. Hubbell [...] and their associates and successors, are hereby created a body corporate and politic in the District of Columbia.

Name of the corporation. Powers.

Sec. 2. That the name of this corporation shall be "The American National Red Cross," and by that name it shall have perpetual succession, with the power to sue and be sued in courts of law and equity within the jurisdiction of the United States; to have and to hold such real and personal estate as shall be convenient and necessary to carry out the purposes of this corporation hereinafter set forth, such real estate to be limited to such quantity as may be necessary for officials use of office buildings; to adopt a seal and the same to alter and destroy at pleasure; and to have the right to have and to use, in carrying out its purposes hereinafter designated, as an emblem and a badge, a Greek red cross on a white ground, as the same has been described in the treaty of Geneva, August twenty-second, eighteen hundred and sixty four, and adopted by the several nations acceding thereto; to ordain and establish by-laws and regulations not inconsistent with laws of the United States of America or any State thereof, and generally to do all such acts and things as may be necessary to carry into effect the provisions of this Act and promote the purposes of said organization; and the corporation is hereby created and designated as the organization which is authorized to act in matters of relief under said treaty. In accordance with article seven, of the treaty, the delivery of the brassard allowed for individuals neutralized in time of war shall be left to military authority.

Purposes.

Sec. 3. That the purposes of this corporation are and shall be

First. To furnish volunteer aid to the sick and wounded of armies in time of war, in accordance with the spirit and conditions of the conference of Geneva of October, eighteen hundred and sixty-three, and also of the treaty of the Red Cross, or the treaty of Geneva of August twenty-second, eighteen hundred and sixty four, to which the United States of America gave its adhesion on March first, eighteen hundred and eighty-two.

Second. And for said purposes to perform all such duties devolved upon a national society by each nation which has acceded to said treaty.

Third. To succeed to all the rights and property which have been hitherto held and to all duties which have heretofore been performed by the American National Red Cross as a corporation duly organized and existing under the laws of the United States relating to the District of Columbia, which organization is hereby dissolved.

Fourth. To act in matters of voluntary relief and in accordance with the military and naval authorities as a medium of communication between the people of the United States of America and their armies, and to act in such matters between similar national societies of other governments through the "Comite International de Secours" and the Government and the people and the armies of the United States of America

Fifth. And to continue and carry on a system of national and international relief in time of peace and apply the same in mitigating the sufferings caused by pestilence, famine, fires, floods, and other great national calamities.

Sixth. And to devise and carry on measures for preventing the same, and generally to promote measures of humanity and the welfare of mankind.

Use of Insignia, etc. Forbidden. Penalty.

Sec. 4. That from and after the passage of this Act it shall be unlawful for any person within the jurisdiction of the United States to falsely and fraudulently hold himself out as, or represent or pretend himself to be a member of or an agent for the American National Red Cross for the purposes of soliciting, collecting, or receiving money or material; or for any person to wear or display the sign of the red cross, or any insignia colored in imitation thereof, for the fraudulent purpose of inducing the belief that he is a member or an agent for the American National Red Cross.

If any person violates the provisions of this section he shall be guilty of a misdemeanor, and shall be liable to a fine of not less than one or not more than five hundred dollars, or imprisonment for a term not exceeding one year, or both, for each and veery offense. The fine so collected shall be paid to the American National Red Cross. The appointment of the chief medical officer shall not be made without the approval in writing of the Secretary of War.

Reports. Amendment.

Sec. 5. That the said American National Red Cross shall, on the first day of January of each year, make and transmit to Congress a full, complete, and itemized report of all receipts and expenditures of whatever kind, and of its proceedings during the preceding year, and shall also give such information concerning its transactions and affairs as the Secretary of State may from time to time require, and, in respect of all business and proceedings in which it may be concerned in connection with the War and Navy Departments of the Government, shall make reports to the Secretary of War and to the Secretary of the Navy, respectively.

Sec. 6. That Congress shall have the right to repeal, alter, or amend this Act at any time.

Approved, June 6, 1900.

Source: Kevin Kosar. March 15, 2006. *The Congressional Charter of the American Red Cross: Overview, History and Analysis.* Congressional Research Service: The Library of Congress, pp. 14-26.

[W]hether the defect was of such a nature that it naturally would produce the nuisance supposed

---

Oliver Wendell Holmes, Jr.
DAVIS V. RICH

62 N.E. 375, 180 Mass. 235 (1902)
Supreme Judicial Court of Massachusetts, Suffolk

Introduction

The law of torts deal with civil wrongs, other than breach of contract, that result in loss or harm to an innocent party, who is entitled to bring a civil suit for damages. Legal scholars have authored many tomes on the subject. Tort law is important because it effects the degree to which courts will hold individuals responsible for actions that might affect others.

In this case, a plaintiff sued after falling on ice on a sidewalk, which was attributed to a leaking, or overflowing pipe, about which a homeowner should have known. Other than requiring the plaintiff to have exercised due care, this decision by Oliver Wendell Holmes, who would later serve on the U.S. Supreme Court upheld the idea of strict liability. This principle is often traced to a British decision in *Fletcher v. Rylands*, 159 Eng. Rep. 737 (Ex. 18965), rev'd, an English case ultimately decided in 1868 by Britain's highest court (the Exchequer Chamber) in which it decided that an individual who built a reservoir was responsible for flooding mine shafts in a coal mine that made them inoperable.

Shugerman (2000) has in turn traced the embrace of strict liability in U.S. law to a series of disasters. The most notable of these was the Johnston Flood of 1889, in which an elite group of U.S. entrepreneurs had escaped liability for not properly inspecting and maintaining a reservoir that burst and resulted in the destruction of a city and more than 2000 deaths.

This case thus serves as a reminder that civil laws and rulings may sometimes have an important impact on the caution exercised by those whose property has the potential for adversely affecting others.

On March 25, 1911, a deadly fire in New York at the Triangle Shirt Waist Factory, which resulted in the deaths of 146 workers (mostly immigrants), led to worker's legislation, including the adoption of worker's compensation laws in many states. Prior to this time, common law doctrines of contributory negligence (employer was not responsible if worker contributed to the injury), the "fellow servant" rule (employer was not responsible if fellow workers were at fault), and the assumption of risk (exempted employers when employees assumed known risks), had made it difficult for workers to win suits for injuries sustained on the job (See Guyton, 1999)

HOLMES, C. J.

This is an action for personal injuries caused by slipping upon a ridge of ice on a sidewalk and falling. There was evidence that the alleged ridge of ice was formed by the freezing of water from a pipe leading from a gutter to a sewer, that the water escaped through a hole in the pipe, or overflowed from the gutter and ran down the pipe on the outside, or in some way was discharged by the pipe upon the sidewalk. The pipe was in the defendant's control. The plaintiff had a verdict, and the case is here on the defendant's exceptions to rulings and refusals to rule.

The plaintiff requested the following instructions: '2. It is the duty of the defendant to so construct and maintain his conductors and spouts as to prevent the water coming therefrom from flowing upon the sidewalk.' '4. It is the duty of the defendant to so construct, manage and care for his premises as to prevent water which accumulates thereon from becoming, in the natural course of events, a source of danger to travellers lawfully using the highway.' '11. If the spout was so maintained on January 31, 1898, as to cause the water to flow upon the sidewalk instead of into the sewer then the defendant is liable without further proof of negligence, if the plaintiff was in the exercise of due care.'

It is stated that the court gave these instructions, adding as to the eleventh that he gave it with the qualification he had already given. And although the qualification referred to was attached specifically only to the eleventh, it is manifest on reading the portions of the charge which are printed that the other rulings also were not left in vacuo as naked, abstract propositions, but were given subject to other requirements repeatedly stated by the judge. Even taking the instructions by themselves, it is not clear that they mean anything more than a statement of the defendant's duty when he knows the conditions. The words 'in the natural course of events' in number 4, and 'maintained' in number 11 point to that interpretation and do not suggest a requirement that the defendant should know at once and at his peril any leak that the spout might spring. Indeed the eleventh request may perhaps be taken to refer to maintaining the pipe so as to discharge from its mouth upon the sidewalk, by deliberate arrangement, as one witness said that it had done. When the defendant does know the conditions, he is bound at his peril to prevent the result.... But the course of the charge should be considered before deciding whether the defendant suffered any wrong. We therefore turn to that.

After explaining the requirement of due care on the plaintiff's part, the judge stated as the next thing to be proved in order to charge the defendant, that there was a nuisance on the sidewalk, and that that nuisance was due to the neglect of the defendant to do what he reasonably could have done to prevent it. Then in more specific form he stated that if the jury found that the ice was a nuisance the question was whether it was produced by water coming from the spout, and then that it must have been a nuisance produced by some defect in the spout or in the gutter near the head of the spout such that water 'would not naturally be carried off.'

The meaning of this is perfectly plain. It was left to the jury to find whether there was a nuisance, and if so whether it was due to a defect, and if so whether the defect was of such a nature that it naturally would produce the nuisance supposed. The word 'naturally' imported that the result must be one which manifestly would come to pass according to common experience if the defect was allowed to remain. A defect in an outside gutter or spout which naturally would produce a nuisance necessarily would be of a cerain magnitude and would be visible on inspection.

The judge then went on to say that the defendant was bound to construct the spout in the first place and to maintain it afterwards in such a manner as to carry off the water so as not to conduct it upon the sidewalk and to create a nuisance. These words, like the requests, probably referred to the defendant's liability in cases where his knowledge of the actual condition of the pipe or his duty to know it would be indisputable. The original plan of construction he would be assumed to know if he constructed the pipe, and the word 'maintain' conveys the notion of such continuance of conditions and mode of working as all would agree that the defendant would be bound to know. There is no doubt that when a man knows that his spout will create a nuisance he is answerable for it.... When knowledge of the damage done, or threatened to the public is established, the strict rule of *Rylands v. Fletcher,* L. R. 3 H. L. 330, is not in question. The general principles of tort established liability unless some special privilege can be made out. We may as well mention in this connection a special exception to a statement by the judge that the defendant could not maintain the conductor in such a manner as naturally to carry water on to the sidewalk so as to create a nuisance. Manifestly this refers to a maintenance with knowledge of the conductor's way of working, and needs no further remark.

Perhaps, however, the instructions held the defendant somewhat more strictly than we have supposed thus far, and meant that he was bound at his peril to find out the defect in his spout within a reasonable time. No particular attention was called to this precise question, but as in other respects the defendant's liability was confined to what a reasonable man could have anticipated, and as it was based in terms upon a neglect to do what the defendant reasonably could to prevent the injury, the meaning probably was not that the defendant was bound to know of a defect at once.

Moreover there does not seem to have been any suggestion that the defect had come suddenly and lately into being, or any ground for such a suggestion. If there was a defect, it probably was an old one.

When we are dealing with a permanent, visible defect that according to common experience will create a nuisance in a highway, undoubtedly the law would require the person responsible for it to find it out, at least within a reasonable time.... The charge of the judge, as we have summed it up, was no stricter than the law as we have stated it, and meant no more. When at the end he read the instructions which we have quoted and said that he had given them substantially, the eleventh with the qualification referred to, obviously no new or contradictory principles were introduced.

The defendant asked for a ruling that the defendant was 'not liable unless he failed to exercise that degree of care in originally building and subsequently maintaining the gutter and conductor which every person of common prudence and caution ordinarily takes of his own concerns.' As to this it is enough to say that if the defendant knew that his gutter or spout created a nuisance, he was absolutely answerable for the harm it did until he stopped it. The request is not discriminating enough to raise a nice question as to precisely how soon the defendant was bound to know of the defect,- whether only within a reasonable time or at once. All that the defendant was entitled to was embraced in the instructions given. The same may be said about the request numbered 13, and also that if there was a hole or leak there was nothing to indicate that it was recent, or that the defendant had shown due care. That was not the defendant's case. His case was that there were no such holes or leaks.

The only other exception was to a refusal to rule that the defendant was not liable unless the jury found that there was a hole in the spout at the time of the accident, from which water escaped and formed an accumulation of ice on the sidewalk. This instruction was given in substance, except that instead of a hole in the spout the jury were allowed to find a defect in the spout or in the gutter near the head of the spout. The larger latitude allowed was right. The plaintiff, it is true, testified to a hole in the spout. But another witness gave evidence which indicated an overflow from the gutter on the outside of the spout, and yet another testified that the spout discharged upon the sidewalk. The defendant's evidence of course must be disregarded as the jury might have disbelieved it.

Exceptions overruled.

Police officers have been authorized by me to KILL any and all persons found engaged in Looting

---

Mayor E.E. Schmitz of San Francisco PROCLAMATION

April 18, 1906

Introduction

Disasters open opportunities for looting, reports of which are often exaggerated, during crisis times when rumors often take the place of hard facts (Sun, 2011). After the catastrophic San Francisco earthquake of 1906, Mayor Schmitz issued the following order, which, despite his reference to "Federal Troops," applied only to local officials and to national guard troops under his command (recall the Posse Comitatus Act of 1878).

Proclamation by the Mayor

The Federal Troops, the members of the Regular Police Force and all Special Police Officers have been authorized by me to KILL any and all persons found engaged in Looting or in the Commission of Any Other Crime.

I have directed all the Gas and Electric Lighting Co.'s not to turn on Gas or Electricity until I order them to do so. You may therefore expect the city to remain in darkness for an indefinite time.

I request all citizens to remain at home from darkness until daylight every night until order is restored.

I WARN all Citizens of the danger of fire from Damaged or Destroyed Chimneys, Broken or Leaking Gas Pipes or Fixtures, or any like cause.

Source: http://www.sfmuseum.org/photos14/killproc.jpg. Accessed 11/16/2016.

I do not consider it paternalism to aid a citizen in these stupendous, unforeseen, and unavoidable calamities

---

William Borah
SPEECH TO THE U.S. SENATE
(1914)

Introduction

William Borah (1865-1940) was a progressive Republican from Idaho who served from 1907 to 1940. In this speech before the Senate, Borah cited Supreme Court decisions and other precedents for using the taxing power to mitigate the effects of natural disasters, in this case by providing bonds to support the price of cotton. In doing so, he sought to confront arguments that such actions were paternalistic and that they would create undue governmental dependency.

I can imagine no wider scope of the power of taxation than that which the Supreme Court of the United States has established. It has said that whatever conserves the general welfare, whatever seems essential to take care of the citizen, to encourage and nourish the spirit of patriotism, or in any way to defend those fundamental principles of society which seem essential to a republican form of government, may be sustained an protected by taxation.... [*Borah cites U.S. v. Gettysburg Electric Railway Co., which upheld the right of eminent domain.*]

Mr. President "peace hath her victories no less renowned than those of war"; and it is just as essential that the Government of the United States protect her citizenship against famine, against impending catastrophies, against poverty superinduced by uncontrollable cause, and against those things which inevitably flow from that condition of affairs—crime and the disintegration and the demoralization of society—as it is to mark the graves of those who have preserved its institutions. There must be soldiers with the capacity and the courage to fight before there are soldiers to bury; there must be a government strong enough and just enough to arouse the love and fealty of the citizen before he will make a good soldier....

If these great people of the South, by no fault of theirs and by reason of a calamity brought on by war, are now facing, as I believe they are, a condition where their homes are menaced, where thousands, if not millions, of poor in this coming winter season will have to depend upon society when society will not be prepared to take care of them, if civil institutions are thus imperiled, it is entitled to be considered as among the duties of Government to bridge the crisis. ... [Borah again quotes from Supreme Court decisions respecting the taxing power]

The Senator from Nebraska [MR. NORRIS] a few moments ago painted a scene with which all western men are familiar – one of those experiences which visited us at a time when the earth had refused to yield and drought and famine had practically swept over our land – and the Senator was of the opinion that the fact that under those conditions no appeal was made might well serve as a precedent against a recurring condition in another part of the country. But, Mr. President, that is not the history of the Anglo Saxon race. We are a progressive people, and we meet emergencies as they arise, provided our experience and our wisdom have been sufficiently exercised and fed to enable us to conceive of a remedy. The people of the United States believe that this Government which they formed is sufficient and efficient, both in its general powers and its reserved powers, to take care of the citizenship of those upon whom it must rest. Because in times past we have not exercised the full powers of government and have not protected our people from suffering and want I do not believe that condition of affairs should be permitted to continue. If we have the power, if the Constitution gives us the right, then it is simply a question of the wisdom of Congress when the necessity of exercising that power shall arise. Modern government must meet modern conditions, and, in my judgment, it can do so by doing no more than applying sound principles of constitutional law to these modern conditions.

Now, Mr. President, we have the power, but should we exercise it? Is this a situation sufficiently momentous and far reaching and of such a nature as to call into exercise this power? Is it a condition that can be dealt with by private efforts, or should it be taken in hand by the combined power of society as represented in the Government? Not as a permanent proposition, but to meet an emergency. All would concede that if it were not an emergency, a crisis affecting millions of people, no such thing as this should be contemplated or considered. All would concede further that the subject should be dealt with in terms as an emergency, as a crisis, and anything we do should be upon the theory that with the going of just and sane powers of government. Paternalism grows space among a feeble, impoverished, and discouraged people. I do not consider it paternalism to aid a citizen in these stupendous, unforeseen, and unavoidable calamities, against which it is impossible for the individual citizens to guard or protect themselves and with which no less than organized society can deal. Now does this condition appeal to us from that standpoint?: It does to me; and I will never hesitate to invoke and use the great power in an emergency any more than I would refuse to use such power to infringe upon the ordinary duties and responsibilities of private endeavor in the ordinary affairs of life.

There is no way by which you can bring the love of these people back to this old Republic – and it sometimes seems essential enough in these days – there is no way by which you can bring the confidence of our people back to these institutions of ours so quickly as to demonstrate to the people of this country that when a great calamity arises the flexibility and adaptability of our Government is such that we may meet that calamity, and one portion of this Nation may share its riches and the comforts with those who have been visited with misfortune. A proud and self-governing people will not be misled or seduced into paternalism by the fact that their Government discloses both the capacity and the willingness to deal with unexpected and unavoidable catastrophes.

I would not support this measure if in my judgment the Constitution of the United States did not justify it; but I am one of those who believe that as from year to year we unravel the powers of this great instrument we shall find that there is very little which inures to the happiness and prosperity and contentment of the hole people which we may not do under this marvelous charter which was given to us a little over 200 years ago. It is not by reason of the unwisdom of our Constitution and the Government under which we live. It is by reason of our failure to exercise those powers in a wise and efficient way, that our people suffer, if they suffer at all.

We are at this time building a railroad in Alaska, loaning the Government in an indirect way to the development of that region of the country. We do this because of the peculiar situation and circumstances and extraordinary conditions which surround private citizens in that country. We have spent millions of dollars to take care of the situation after great fires in some of our cities; after the earthquake at San Francisco; and we have sent large sums of money to sufferers in foreign lands by reason of earthquakes and other disasters. Shall we hesitate, or shall we find our Government incapacitated to deal with extraordinary conditions in our own country when we have dealt with them when they have developed in other countries? We spend millions of dollars year after year to protect the farms and homes along the Mississippi River. We loan the credit of the Government yearly and constantly to our private banks, organized for private gain and purely personal interests. Originally the banks did not ask for any more than the power to issue their notes, which should serve as currency – a most extraordinary thing, indeed – but now they not only have the privilege to issue the notes which shall circulate as our currency, but the Government of the United States guarantees these notes, which are nothing more than the notes of private corporations organized for gain. We loaned the credit of the Government, in addition to donating millions of acres of land, to the building of railroads, and took mortgages for our security. We have, as I have said, loaned the credit of the Government to the western settler to aid him in the building of his home and do not think the western settler has lost any of his initiative, his self-reliance, and his good citizenship. He knows the Government is but aiding him to do that which private enterprise will not or can not do, and that as soon as the desert is conque4red the citizen is expected, and is perfectly willing, to pay the Government and relieve himself of all obligations to or dependence upon the Government.

There is no difficulty about the powers of our Government. It is sufficient and efficient to do justice to all men and to permit the most extended development and progress of the human family....

Source: 51 Cong. Rec. HR18891, 16777-16777 (daily ed. October 17, 1914).

[S]o unusual and unprecedented in its nature as to amount in law to an act of God

---

CHICAGO R.I. & P. RY. CO. V. MORTON

1916 OK 552, 157 P. 917, 57 Okla.766, (1916)

Introduction

It is sometimes difficult to disentangle the degree to which human decisions aggravate natural disasters, which insurance companies and governments often refer to "acts of God." In this Oklahoma case, a plaintiff contended that the construction of a railroad bridge was done negligently in that it backed up a creek that overflowed on the plaintiff's property and caused crop damage. In reviewing the conviction, the Supreme Court of Oklahoma upheld the following instructions, which the lower court judge had issued to the jury.

RUMMONS, C.

"You are further instructed that an 'act of God,' such as an unprecedented rainfall and resulting flood, which will excuse from liability, must not only be the proximate cause of the loss, but it must be the sole cause. If, however, the injury is caused by an act of God, commingled with the negligence of the defendant as an efficient and contributing cause, and the injury would not have occurred except for such negligence, the defendant would be liable.

"And you are further instructed that whether the flood of August 5, 1911, which plaintiff claims damaged his property, was an ordinary flood or an extraordinary flood, so unusual and unprecedented in its nature as to amount in law to an act of God, thus relieving the defendant of all liability, are questions for the jury to decide; and, if you find from the evidence that the flood in question was so unusual and unprecedented that its extent and resulting effects could not have been reasonably anticipated and provided against by an ordinarily careful person in defendant's situation, then the defendant is not liable in damages caused thereby, unless you further find from the evidence that, even though such flood was in law an act of God, such injury would not have occurred except for the negligence of the defendant in maintaining the bridges across Driftwood creek, and the roadbed adjacent thereto, in such a condition as to obstruct the natural flow of the water in time of ordinary floods, and that such negligence was an efficient and contributing concurrent cause of the injury, then and in that event the defendant is liable for such injuries as may have been proven in the case."

Associations organized ... exclusively for religious, charitable, scientific, or educational purposes

---

WAR REVENUE ACT OF 1917

Introduction

The Sixteenth Amendment, which states ratified in 1913, provided the constitutional authority for an income tax, which the Supreme Court had previously invalidated. Income taxes were especially critical during times of war, which often disrupted customs taxes. The Revenue Act of 1917, which Congress adopted during World War I, is an example of the manner in which the government sought to encourage aid to charitable organizations, which were often on the front lines of disaster relief, by exempting such contributions from this tax. Such deductions continue to be a feature of contemporary tax policies, and churches and private charities continue to provide workers and supplies in the aftermath of disasters.

[Exemptions shall be granted for] contributions or gifts actually made within the year to corporations or associations organized and operated exclusively for religious, charitable, scientific, or educational purposes, or to societies for the prevention of cruelty to children or animals, no part of the net income of which inures to the benefit of any private stockholder or individual, to an amount not in excess of fifteen per centum of the taxpayer's taxable net income as computed without the benefit of this paragraph. Such contributions or gifts shall be allowable as deductions only if verified under rules and regulations prescribed by the Commissioner of Internal Revenue, with the approval of the Secretary of the Treasury.

Source: War Income Tax Revenue Act of 1917, ch. 63, 1201(2), reprinted in J.S. Seldman, Seidman's *Legislative History of Federal Income Tax Laws, 1928-1861,* at 944 (1938).

[T]o save life and prevent suffering, and not to replace losses

---

SPECIAL REGULATIONS NO. 67, REGULATIONS GOVERNING FLOOD RELIEF WORK OF THE WAR DEPARTMENT

1917

Introduction

The resources of the armed forces are significantly larger than those of FEMA and of private agencies, and troops often outnumber their employees in the field. The following regulations were issued to govern such relief efforts before the latter organizations were even established. These regulations put great emphasis on encouraging locals to help themselves and on avoiding future dependency upon governmental help.

WAR DEPARTMENT

Washington, October 12, 1917.

The following regulations governing relief work of the War Department in connection with flood, fire, earthquake, or other great catastrophe are published to the Army for the information and guidance of all concerned.

BY THE ORDER OF THE SECRETARY OF WAR:

TASKER H. BLISS,

General, Chief of Staff.

1. **Authority for undertaking relief work.**—It is not contemplated that relief work will be undertaken by the War Department in connection with flood, fire, earthquake, or other great catastrophe without specific authority of Congress, unless the overruling demands of humanity compel immediate action to prevent starvation and extreme suffering and local resources are clearly inadequate to cope with the situation.

2. Reports of catastrophe to The Adjutant General of the Army and department commander.—In times of great catastrophe the senior officer present for duty in the affected district will make an investigation as soon as practicable and report by wire to The Adjutant General of the Army the nature and extent of the catastrophe, the area involved, the number of people seriously affected and their needs, the measures being taken by local authorities, and whether the suffering is, or probably will be, beyond the powers of State or local resources. These reports shall be furnished at the same time and in the same way to the commanding general of the territorial department within the limits of which the affected area is located. In cases of flood the investigation will be made and the reports forwarded by the district Engineer officers whose districts are affected. Whenever a flood of dangerous proportions is foreseen, district Engineer officers will keep The Adjutant General of the Army and the territorial department commander informed by wire of developments.

28. Relief limited to those entitled to it.—Care should be taken to see that aid is extended solely to those actually in need of relief, and it must be ascertained, not only that the need really exists, but that the needy condition is due to the flood, fire, or other catastrophe which made Government aid necessary. It must also be conclusively shown that the applicant for aid has no other means of securing assistance....

29. Reimbursement for losses not authorized.—In no case will Government supplies or funds be used to reimburse sufferers for losses sustained. The Army's mission in the territory affected is to save life and prevent suffering, and not to replace losses.

30. Sale of supplies.—If the catastrophe be of such a character that no supplies are available except those in possession of the military authorities, such supplies may be sold at cost to those who are able to pay for same, but such sales must be made under the restrictions heretofore laid down that the supplies are needed to prevent suffering directly resulting from the catastrophe. The funds received from this source will be deposited with the Treasurer of the United States in accordance with regulations.

44. Methods of providing employment for beneficiaries.— Wherever practicable, beneficiaries should be required to earn the aid given them, either by services on Government work or work of local benefit, provided no employment on pay for private firms or persons is procurable. A labor bureau should be established whenever possible to find employment for the idle. To this bureau all applicants for aid should be required to report regularly as an indispensable condition to securing supplies. Able-bodies applicants who are offered remunerative work should be dropped from the list of beneficiaries to whom supplies are being issued, whether the offer of work is accepted or rejected. The establishment of these labor bureaus comes very properly within the sphere of activity of local relief committees, is an important adjunct of relief work, and a safeguard against a promiscuous distribution of supplies and consequent demoralization of labor conditions thereby. If Government or municipal work is provided in the endeavor to furnish something for the idle to do, it should not be of such a character or so conducted as to encourage persons so employed to depend upon Federal aid beyond the point of actual necessity. Relief should be given only for such time as is absolutely necessary to permit sufferers to readjust their affairs and again become self-supporting.

Source: Taken from Special Regulations No. 67, *Regulations Governing Flood Relief Work of the War Department.* Washington: Government Printing Office: 1917.

For controlling the floods and for the general improvement of the Mississippi River

---

FLOOD CONTROL ACT

1917

Introduction

Although the Mississippi and Sacramento Rivers had long been subject to floods, the Progressive Era led to hopes that technology might control such events. This act, which created a model for future federal/state cooperation, chiefly concentrated on the construction of levees to mitigate future disasters. In time, the government began to realize that such measures were often insufficient to contain the fury of nature and that they sometimes alleviated flooding in some areas only to increase them for others, especially those downstream.

Chap. 144. – An Act to provide for the control of the floods of the Mississippi River and of the Sacramento River, Calif., and for other purposes.

Be it enacted by the Senate and House of Representatives of the United States of America in Congress assembled, That for controlling the floods of the Mississippi River and continuing its improvement from the Head of Passes and the mouth of the Ohio River the Secretary of War is hereby empowered, authorized, and directed to carry on, continuously, by hired labor or otherwise, the plans of the Mississippi River Commission heretofore or hereafter adopted; to be paid for as appropriations may from time to time be made by law, not to exceed in the aggregate $45,000,000: Provided, That not more than $10,000,000 shall be expended therefore during any one fiscal year.

(a) All money appropriated under authority of this section shall be expended under the direction of the Secretary of War in accordance with the plans, specifications, and recommendations of the Mississippi River Commission as approved by the Chief of Engineers, for controlling the floods and for the general improvement of the Mississippi River, and for surveys, including the survey from the Head of the Passes to the headwaters of the river, and a survey of the Atchafalaya Outlet so far as may be necessary to determine the cost of protecting its basin from the flood waters of the Mississippi River either by divorcement from the Mississippi River or by other means, and for salaries, clerical, office, traveling, and miscellaneous expenses of the Mississippi River Commission.

(b) That no money appropriated under authority of this section shall be expended in the construction or repair of any levee unless and until assurances have been given satisfactory to the commission that local interests protected thereby will contribute for such construction and repair a sum which the commission shall determine to be just and equitable but which shall not be less than one-half of such sum as may have been allotted by the commission for such work: Provided, That such contribution s shall be expended under the direction of the commission, or in such manner as it may require or approve, but no contribution mae by any State or levee district shall be expended in any other State or levee district except with the approval of the authorities of the State or district so contributing

(c) [Omitted]

(d) No money appropriated under authority of this Act shall be expended in payment for any right of way for any levee which may be constructed in cooperation with any State or levee district under authority of this Act, but all such rights of way shall be provided free of cost to the United States: Provided, That no money paid or expense incurred by any State or levee District in securing such rights of way, or in any temporary works of emergency during an impending flood, or for the maintenance of any levee line, shall be computed as part of the contribution of such State or levee district toward the construction or repair of any levee within the meaning of paragraph (b) of this section.

[Other provisions omitted.]

Source: Sixty-Fourth Congress, Sess. 22. Ch. 144.

PART V

The New Deal and World War II

[T]he present economic depression has created a serious national emergency

---

THE FEDERAL EMERGENCY RELIEF ACT OF 1933

May 12, 1933

Introduction

Although governments had a long history of intervening to battle natural disasters, opinion was far more divided on the role that governments should play in battling unemployment and economic depression. After the stock market crash of 1929 plunged the nation into the Great Depression, Franklin D. Roosevelt ran in 1932 on a program that stressed aggressive national programs, as part of a New Deal to combat such evils. One of Roosevelt's key proposals was the creation of the Federal Emergency Relief Administration (FERA), which treated high unemployment much as the government had previously responded to natural disasters.

AN ACT

To provide for cooperation by the Federal Government with the several States and Territories and the District of Columbia in relieving the hardship and suffering caused by unemployment, and for other purposes.

Be it enacted by the Senate and House of Representatives of the United States of America in Congress assembled, That the Congress hereby declares that the present economic depression has created a serious emergency, due to widespread unemployment and increasing inadequacy of State and local relief funds, resulting in the existing or threatened deprivation of a considerable number of families and individuals of the necessities of life, and making it imperative that the Federal Government cooperate more effectively with the several States and Territories and the District of Columbia in furnishing relief to their needy and distressed people.

Sec. 2. (a) The Reconstruction Finance Corporation is authorized and directed to make available out of the funds of the Corporation not to exceed $500,000,000, in addition to the funds authorized under title I of the Emergency Relief and Construction Act of 1932, for expenditure under the provisions of this Act upon certification by the Federal Emergency Relief Administrative provided for in section 3.

(b) The amount of notes, debentures, bonds, or other such obligations which the Reconstruction Finance Corporation is authorized and empowered under section 9 of the Reconstruction Finance Corporation Act, as amended, to have outstanding at any one time is increased by $500,000,000: Provided, That no such additional notes, debentures, bonds, or other such obligations authorized by this subsection shall be issued except at such times and in such amounts as the President shall approve.

(c) After the expiration of ten days after the date upon which the Federal Emergency Relief Administrator has qualified and has taken office, no application shall be approved by the Reconstruction Finance Corporation under the provisions of title I of the Emergency Relief and Construction Act of 1932, and the Federal Emergency Relief Administrator shall have access to all files and records of the Reconstruction Finance Corporation relating to the administration of funds under title I of such Act. At the expiration of such ten-day period, the unexpended and unobligated balance of the funds authorized under title I of such Act shall be available for the purposes of this Act.

Sec. 3. (a) There is hereby created a Federal Emergency Relief Administration, all the powers of which shall be exercised by a Federal Emergency Relief Administrator (referred to in this Act as the *Administrator*) to be appointed by the President, by and with the advice and consent of the Senate. The Administrator shall receive a salary to be fixed by the President at not to exceed $10,000, and necessary traveling and subsistence expenses within the limitations prescribed by law for civilian employees in the executive branch of the Government. The Federal Emergency

Relief Administration and the office of Federal Emergency Relief Administrator shall case to exist upon the expiration of two years after the date of enactment of this Act, and the unexpended balance on such date of any funds made available under the provisions of this Act shall be disposed of as the Congress may by law provide.

(b) The Administrator may appoint and fix the compensation of such experts and their appointment may be made and compensation fixed without regard to the civil service laws, or the Classification Act of 1923, as amended...

(c) In executing any of the provisions of this Act, the Administrator, and any person duly authorized or designated by him, may conduct any investigation pertinent or material to the furtherance of the purposes of this Act and, at the request of the President, shall make such further investigations and 0studies as the President may deem necessary in dealing with problems of unemployment relief.

(d) omitted

Sec. 4. (a) Out of the funds of the Reconstruction Finance Corporation made available by this Act, the Administrator is authorized to make grants to the several States to aid in meeting the costs of furnishing relief and work relief and in relieving the hardship and suffering caused by unemployment in the form of money, service, materials, and/or commodities to provide the necessities of life to persons in need as a result of the present emergency, and/or to their dependents, whether resident, transient, or homeless.

(b) Of the amounts made available by this Act not to exceed $250,000,000 shall be granted to the several States applying therefor, in the following manner: Each State shall be entitled to receive grants equal to one third of the amount expended by such State, including the civil subdivisions thereof, out of public moneys from all sources for the purposes set forth in subsection (a) of this section; and such grants shall be made quarterly, beginning with the second quarter in the calendar year 1933, and shall be made during any quarter upon the basis of such expenditures certified by the States to have been made during the preceding quarter.

c-f and Sections 5-7 omitted

Sec. 8. This Act may be cited as the *Federal Emergency Relief Act of 1933.*

Approved, May 12, 1933

Source: United States Statutes at Large (73rd Cong., Sess. I, Chp. 30, p. 55-58).

Source: Found at the Social Welfare History Project, VCU Libraries. http://socialwelfare.library.vcu.edu/eras/great-depression;/federal-emergency-relief-act-of-1... Accessed 12/3/2016.

To improve the navigability and to provide for the flood control of the Tennessee River

---

TENNESSEE VALLEY AUTHORITY ACT

1933

Introduction

Among the many programs that Franklin D. Roosevelt introduced, one of the most consequential, and controversial, was the creation of the Tennessee Valley Authority. Long advocated by Senator George W. Norris of Nebraska, this law created a corporation to build a massive system of dams and power plants to control flooding and generate power for the Tennessee Valley, which was one of the most economically-depressed areas of the country. The Supreme Court upheld the constitutionality of this act in *Ashwander v. Tennessee Valley Authority,* 297 U.S. 288 (1936), where justices also developed elaborate rules of judicial restraint. Critics charged that the law, which often required the displacement of entire towns, whose lands were flooded by dams, usurped functions that were more appropriate for private enterprises, but the law clearly resulted both in better flood control and in economic development for the area.

TENNESSEE VALLEY AUTHORITY ACT

AN ACT

To improve the navigability and to provide for the flood control of the Tennessee River; to provide for reforestation and the proper use of marginal lands in the Tennessee Valley; to provide for the agricultural and industrial development of said valley; to provide for the national defense by the creation of a corporation for the operation of Government properties at and near Muscle Shoals in the State of Alabama, and for other purposes.

Be it enacted by the Senate and House of Representatives of the United States of America in Congress assembled, That for the purpose of maintaining and operating the properties now owned by the United States in the vicinity of Muscle Shoals, Alabama, in the interest of the national defense and for agricultural and industrial development, and to improve navigation in the Tennessee River and Mississippi River Basins, there is hereby created a body corporate by the name of the "Tennessee Valley Authority" (hereinafter referred to as the "Corporation"). The Board of Directors first appointed shall be deemed the incorporator, and the incorporation shall be held to have been effected from the date of the first meeting of the board. The Act may be cited as the "Tennessee Valley Authority Act of 1933."

To preserve natural resources [and] control floods

SOIL CONSERVATION ACT

Pub. L. 74-76, April 27, 1935

Introduction

The 1930s witnessed drought that led to dust storms throughout the prairies of North America. They resulted in a "Dust Bowl" that drove numerous families from their lands (John Steinbeck's *Grapes of Wrath* described this situation as did the widely-circulated black-and-white photographs of Dorothea Lange) and undermined agricultural productivity. The Soil Conservation Act of 1935 sought to address this problem. After the Supreme Court declared the Agricultural Adjustment Act to be unconstitutional in *United States v. Butler*, Congress subsequently adopted the Soil Conservation and Domestic Allotment Act in 1936, which the Court eventually upheld. It paid farmers to withdraw some arable land from tillage.

AN ACT

To provide for the protection of land resources against soil erosion, and for other purposes.

Be it enacted by the Senate and House of Representatives of the United States of America in Congress assembled, That it is hereby recognized that the wastage of soil and moisture resources on farm, grazing, and forest lands of the Nation, resulting from soil erosion, is a menace to the national welfare and that it is hereby declared to be the policy of Congress to provide permanently for the control and prevention of soil erosion and thereby to preserve natural resources, control floods, prevent impairment of reservoirs, and maintain the navigability of rivers and harbors, protect public health, public lands and relieve unemployment, and the Secretary of Agriculture, from now on, shall coordinate and direct all activities with relation to soil erosion and in order to effectuate this policy is hereby authorized from time to time—

(1) To conduct surveys, investigations, and research relating to the character of soil erosion and the preventive measures needed, to publish the results of any such surveys, investigations, or research, to disseminate information concerning such methods, and to conduct demonstrational projects in areas subject to erosion by wind or water;

(2) To carry out preventive measures, including, but not limited to, engineering operations, methods of cultivation, the growing of vegetation, and changes in use of land...

SEC. 5. The Secretary of Agriculture shall establish an agency to be known as the "Soil Conservation Service," to exercise the powers conferred on him by this Act...

Source: Pub. L. 784-46, 49 Stat. pp. 163-164.

"[T]he power of Congress to authorize expenditure of public moneys for public purposes is not limited by the direct grants of legislative power found in the Constitution"

UNITED STATES V. BUTLER

297 U.S. 1; 56 S. Ct. 312; 80 L. Ed. 477; (1936)

Introduction

At the beginning of the New Deal, President Franklin D. Roosevelt introduced a host of new programs designed to combat the Great Depression. The Agricultural Adjustment Act was among the most far reaching. It sought to provide stability to agricultural prices through a whole series of regulations. Roosevelt sought to justify the program through the congressional power over taxing and spending, the same justification that generally served as the basis for disaster relief. In this decision, the Court officially sided with Justice Joseph Story's expansive, but limited, view of this power. In somewhat of an about face, however, the Court decided that the act at issue was unconstitutional because it interfered with traditional state powers over agriculture, which the court thought to be protected by the Tenth Amendment, which reserved un-enumerated powers to the states. Justice Owen Roberts (1875-1955), who wrote the decision,would eventually form part of the Court majority that would support most New-Deal programs. Justice Harlan Fiske Stone (1872-1946), who penned the dissent (largely focusing on the Tenth Amendment interpretation), would later serve as Chief Justice.

Earlier precedents for the Court's decision include *Field v. Clark*, 143 U.S. 649 (1892) and *United States v. Realty Co.*, 163 U.S. 427 (1896).

MR. JUSTICE ROBERTS delivered the opinion of the Court.

In this case we must determine whether certain provisions of the Agricultural Adjustment Act, 1933, conflict with the Federal Constitution. [In the first part of the opinion, Justice Roberts summarizes the law and decides that the parties had proper standing to bring the case before the Court.]

The clause thought to authorize the legislation, -- the first, confers upon the Congress power "to lay and collect Taxes, Duties, Imposts and Excises, to pay the Debts and provide for the common Defence and general Welfare of the United States...." It is not contended that this provision grants power to regulate agricultural production upon the theory that such legislation would promote the general welfare. The Government concedes that the phrase "to provide for the general welfare" qualifies the power "to lay and collect taxes." The view that the clause grants power to provide for the general welfare, independently of the taxing power, has never been authoritatively accepted. Mr. Justice Story points out that if it were adopted "it is obvious that under color of the generality of the words, to 'provide for the common defence and general welfare,' the government of the United States is, in reality, a government of general and unlimited powers, notwithstanding the subsequent enumeration of specific powers." The true construction undoubtedly is that the only thing granted is the power to tax for the purpose of providing funds for payment of the nation's debts and making provision for the general welfare.

Nevertheless the Government asserts that warrant is found in this clause for the adoption of the Agricultural Adjustment Act. The argument is that Congress may appropriate and authorize the spending of moneys for the "general welfare"; that the phrase should be liberally construed to cover anything conducive to national welfare; that decision as to what will promote such welfare rests with Congress alone, and the courts may not review its determination; and finally that the appropriation under attack was in fact for the general welfare of the United States.

Congress is expressly empowered to lay taxes to provide for the general welfare. Funds in the Treasury as a result of taxation may be expended only through appropriation. (Art. I, § 9, cl. 7.) They can never accomplish the objects for which they were collected unless the power to appropriate is as broad as the power to tax. The necessary implication from the terms of the grant is that the public funds may be appropriated "to provide for the general welfare of the United States." These words cannot be meaningless, else they would not have been used. The conclusion must be that they were intended to limit and define the granted power to raise and to expend money. How shall they be construed to effectuate the intent of the instrument?

Since the foundation of the Nation sharp differences of opinion have persisted as to the true interpretation of the phrase. Madison asserted it amounted to no more than a reference to the other powers enumerated in the subsequent clauses of the same section; that, as the United States is a government of limited and enumerated powers, the grant of power to tax and spend for the general national welfare must be confined to the enumerated legislative fields committed to the Congress. In this view the phrase is mere tautology, for taxation and appropriation are or may be necessary incidents of the exercise of any of the enumerated legislative powers. Hamilton, on the other hand, maintained the clause confers a power separate and distinct from those later enumerated, is not restricted in meaning by the grant of them, and Congress consequently has a substantive power to tax and to appropriate, limited only by the requirement that it shall be exercised to provide for the general welfare of the United States. Each contention has had the support of those whose views are entitled to weight. This court has noticed the question, but has never found it necessary to decide which is the true construction. Mr. Justice Story, in his *Commentaries*, espouses the Hamiltonian position. We shall not review the writings of public men and commentators or discuss the legislative practice. Study of all these leads us to conclude that the reading advocated by Mr. Justice Story is the correct one. While, therefore, the power to tax is not unlimited, its confines are set in the clause which confers it, and not in those of § 8 which bestow and define the legislative powers of the Congress. It results that the power of Congress to authorize expenditure of public moneys for public purposes is not limited by the direct grants of legislative power found in the Constitution.

But the adoption of the broader construction leaves the power to spend subject to limitations. As Story says: "The Constitution was, from its very origin, contemplated to be the frame of a national government, of special and enumerated powers, and not of general and unlimited powers." Again he says: "A power to lay taxes for the common defence and general welfare of the United States is not in common sense a general power. It is limited to those objects. It cannot constitutionally transcend them."

That the qualifying phrase must be given effect all advocates of broad construction admit. Hamilton, in his well known Report on Manufactures, states that the purpose must be "general, and not local." Monroe, an advocate of Hamilton's doctrine, wrote: "Have Congress a right to raise and appropriate the money to any and to every purpose according to their will and pleasure? They certainly have not." Story says that if the tax be not proposed for the common defence or general welfare, but for other objects wholly extraneous, it would be wholly indefensible upon constitutional principles. And he makes it clear that the powers of taxation and appropriation extend only to matters of national, as distinguished from local welfare.

As elsewhere throughout the Constitution the section in question lays down principles which control the use of the power, and does not attempt meticulous or detailed directions. Every presumption is to be indulged in favor of faithful compliance by Congress with the mandates of the fundamental law. Courts are reluctant to adjudge any statute in contravention of them. But, under our frame of government, no other place is provided where the citizen may be heard to urge that the law fails to conform to the limits set upon the use of a granted power. When such a contention comes here we naturally require a showing that by no reasonable possibility can the challenged legislation fall within the wide range of discretion permitted to the Congress. How great is the extent of that range, when the subject is the promotion of the general welfare of the United States, we hardly need remark. But, despite the breadth of the legislative discretion, our duty to hear and to render judgment remains. If the statute plainly violates the stated principle of the Constitution we must so declare.

We are not now required to ascertain the scope of the phrase "general welfare of the United States" or to determine whether an appropriation in aid of agriculture falls within it. Wholly apart from that question, another principle embedded in our Constitution prohibits the enforcement of the Agricultural Adjustment Act. The act invades the reserved rights of the states. It is a statutory plan to regulate and control agricultural production, a matter beyond the powers delegated to the federal government. The tax, the appropriation of the funds raised, and the direction for their disbursement, are but parts of the plan. They are but means to an unconstitutional end.

MR. JUSTICE STONE, dissenting.

I think the judgment should be reversed.

The present stress of widely held and strongly expressed differences of opinion of the wisdom of the Agricultural Adjustment Act makes it important, in the interest of clear thinking and sound result, to emphasize at the outset certain propositions which should have controlling influence in determining the validity of the Act. They are:

1. The power of courts to declare a statute unconstitutional is subject to two guiding principles of decision which ought never to be absent from judicial consciousness. One is that courts are concerned only with the power to enact statutes, not with their wisdom. The other is that while unconstitutional exercise of power by the executive and legislative branches of the government is subject to judicial restraint, the only check upon our own exercise of power is our own sense of self-restraint. For the removal of unwise laws from the statute books appeal lies not to the courts but to the ballot and to the processes of democratic government.

2. The constitutional power of Congress to levy an excise tax upon the processing of agricultural products is not questioned. The present levy is held invalid, not for any want of power in Congress to lay such a tax to defray public expenditures, including those for the general welfare, but because the use to which its proceeds are put is disapproved.

3. As the present depressed state of agriculture is nation wide in its extent and effects, there is no basis for saying that the expenditure of public money in aid of farmers is not within the specifically granted power of Congress to levy taxes to "provide for the ... general welfare." The opinion of the Court does not declare otherwise. ...

MR. JUSTICE BRANDEIS and MR. JUSTICE CARDOZO join in this opinion.

By any Act of Congress appropriating funds for work relief or relief purposes

---

THE HATCH ACT

1939

Introduction

As governments assumed increased responsibility, there were growing fears that those who held office would use their positions to extract support from those under them or from those to whom they were extending governmental largesse. The Hatch Act sought to prevent governmental employees from participating in direct political activities that might cast doubt on their commitment to the common good. The act also attempted to prevent governmental use of relief programs as a way of bolstering party support. As responses to Hurricane Katrina and other disasters have shown, however, this and subsequent laws cannot prevent voters from judging governmental officials by the perceived generosity and effectiveness of the aid that they provide during emergencies.

76TH UNITED STATES CONGRESS
1ST SESSION
An Act
To prevent pernicious political activities.

Be it enacted by the Senate and House of Representatives of the United States of America in Congress assembled,

That it shall be unlawful for any person to intimidate, threaten, or coerce, or to attempt to intimidate, threaten, or coerce, any other person for the purpose of interfering with the right of such other person to vote or to vote as he may choose, or of causing such other person to vote for, or not to vote for, any candidate for the office of

President, Vice President, Presidential elector, Member of the Senate, or Member of the House of Representatives at any election held solely or in part for the purpose of selecting a President, a Vice President, a Presidential elector, or any Member of the Senate or any Member of the House of Representatives, Delegates or Commissioners from the Territories and insular possessions.

SEC. 2.

It shall be unlawful for any person employed in any administrative position by the United States, or by any department, independent agency, or other agency of the United States (including any corporation controlled by the United States or any agency thereof, and any corporation all of the capital stock of which is owned by the United States or any agency thereof), to use his official authority for the purpose of interfering with, or affecting the election or the nomination of any candidate for the office of President, Vice President, Presidential elector, Member of the Senate, or Member of the House of Representatives, Delegates or Commissioners from the Territories and insular possessions.

SEC. 3.

It shall be unlawful for any person, directly or indirectly, to promise any employment, position, work, compensation, or other benefit, provided for or made possible in whole or in part by any Act of Congress, to any person as consideration, favor, or reward for any political activity or for the support of or opposition to any candidate or any political party in any election.

SEC. 4.

Except as may be required by the provisions of subsection (b), section 9 of this Act, it shall be unlawful for any person to deprive, attempt to deprive, or threaten to deprive, by any means, any person of any employment, position, work, compensation, or other benefit provided for or made possible by any Act of Congress appropriating funds for work relief or relief purposes, on account of race, creed, color, or any political activity, support of, or opposition to any candidate or any political party in any election.

SEC. 5.

It shall be unlawful for any person to solicit or receive or be in any manner concerned in soliciting or receiving any assessment, subscription, or contribution for any political purpose whatever from any person known by him to be entitled to or receiving compensation, employment, or other benefit provided for or made possible by any Act of Congress appropriating funds for work relief or relief purposes.

SEC. 6.

It shall be unlawful for any person for political purposes to furnish or to disclose, or to aid or assist in furnishing or disclosing, any list or names of persons receiving compensation, employment, or benefits provided for or made possible by any Act of Congress appropriating, or authorizing the appropriation of, funds for work relief or relief purposes, to a political candidate, committee, campaign manager, or to any person for delivery to a political candidate, committee, or campaign manager, and it shall be unlawful for any person to receive any such list or names for political purposes.

SEC. 7.

No part of any appropriation made by any Act, heretofore or hereafter enacted, making appropriations for work relief, relief, or otherwise to increase employment by providing loans and grants for public-works projects, shall be used for the purpose of, and no authority conferred by any such Act upon any person shall be exercised or administered for the purpose of, interfering with, restraining, or coercing any individual in the exercise of his right to vote at any election.

SEC. 8-11.

[Omiteed]

Approved August 2, 1939.

Source: Wikisource.

By virtue of the authority vested in me
by the Constitution and the statues

Franklin D. Roosevelt
Executive Order 8629
ON THE OFFICE OF PRODUCTION MANAGEMENT AND THE OFFICE FOR EMERGENCY MANAGEMENT

January 7, 1941

Introduction

Early federal responses to emergencies appeared to be largely episodic. As World War II started in Europe, President Roosevelt anticipated that the nation would eventually join the conflict, as it did after the surprise Japanese attack on the American naval base in Pearl Harbor, Hawaii on December 7, 1941. He further recognized that the nation was largely unprepared. In 1939, he accordingly created an Office for Emergency Management within the executive office of the president, and he issued several subsequent orders on the subject.

This follow-up order demonstrates that the federal government often pays greater attention to threats from foreign governments than from natural disasters, but that the efforts to combat them often overlap.

By virtue of the authority vested in me by the Constitution and the statutes, and in order to define further the functions and duties of the Office for Emergency Management with respect to the national emergency as declared by the President to exist on September 8, 1939, and to increase production for the national defense through mobilization of material resources and the industrial facilities of the Nation, it is hereby ordered:

1. There shall be in the Office for Emergency Management of the Executive Office of the President, an Office of Production Management which shall consists of (1) a Director General, and (2) an Associate Director General, each to be appointed by the President, (3) the Secretary of War, and (4) the Secretary of the Navy. The members shall serve as such without compensation but shall be entitled to actual necessary transportation, subsistence, and other expenses incidental to the performance of their duties.

2. With such advice and assistance as it may require from other departments and agencies of the Federal Government, and subject to such regulations or directions as the President may from time to time pre scribe, and subject further to the general policy that the Departments of War and Navy and other departments and agencies of the Government will be utilized to the maximum extent compatible with efficiency, the Office of Production Management shall: [the order proceeds to list these responsibilities.] ...

The Office for Emergency Management shall have the following duties and functions:

(a) To advise and assist the President in the discharge of extraordinary responsibilities imposed upon him by any emergency arising out of war, the threat of war, imminence of war, flood, drought, or other condition threatening the public peace or safety.

(b) To serve as a division of the Executive Office of the President, with such subdivisions as may be required, through which the President, during any emergency, may coordinate and supervise and, in appropriate cases, direct the activities of agencies, public or private, in relation thereto.

(c) To serve as a channel of communication between such agencies and the President concerning emergency activities, to keep the President currently advised of their progress, to assemble and analyze information concerning additional measures that should be taken, and to assist in the preparation of recommendations for any necessary legislation.

(d) To provide and maintain liaison during any such emergency with other divisions of the Executive Office of the President and with other agencies, public or private, for the purpose of bringing about maximum utilization and coordination of their services and facilities.

(e) To advise and assist the President upon or before termination of any such emergency with respect to any measures that may be needful to facilitate a restoration of normal administrative relations and to ameliorate the con-sequences of the emergency.

(f) To perform such other duties and functions with respect to any such emergency as the President may from time to time direct. . . .

Source: The American Presidency Project, http://www.presidency.ucsb.edu/ws/?p;id=16114. Accessed 11/19/2016.

A date which will live in infamy

Franklin D. Roosevelt
ADDRESS TO CONGRESS REQUESTING DECLARATION OF WAR

December 8, 1941

Introduction

Although this book focused chiefly on defense against natural disasters, which typically threaten a single state or region, President Franklin D. Roosevelt confronted a depression that threatened the entire national and world economies. In time, however, Japanese and German aggression posed an even greater threat to his expansive New Deal programs. After the Japanese attack on Pearl Harbor of December 7, 1941, Roosevelt called upon Congress to declare war. After Congress did so, Germany and Italy joined the conflict on the side of Japan. Roosevelt declared that "Dr. Win the War" had replaced "Dr. New Deal," and he devoted far more attention during the rest of his administration to winning the war than on his earlier domestic programs.

Mr. Vice President, and Mr. Speaker, and Members of the Senate and House of Representatives:Yesterday, December 7, 1941 —a date which will live in infamy—the United States of America was suddenly and deliberately attacked by naval and air forces of the Empire of Japan. The United States was at peace with that Nation and, at the solicitation of Japan, was still in conversation with its Government and its Emperor looking toward the maintenance of peace in the Pacific. Indeed, one hour after Japanese air squadrons had commenced bombing in the American Island of Oahu, the Japanese Ambassador to the United States and his colleague delivered to our Secretary of State a formal reply to a recent American message. And while this reply stated that it seemed useless to continue the existing diplomatic negotiations, it contained no threat or hint of war or of armed attack. It will be recorded that the distance of Hawaii from Japan makes it obvious that the attack was deliberately planned many days or even weeks ago. During the intervening time the Japanese Government has deliberately sought to deceive the United States by false statements and expressions of hope for continued peace. The attack yesterday on the Hawaiian Islands has caused severe damage to American naval and military forces. I regret to tell you that very many American lives have been lost. In addition American ships have been reported torpedoed on the high seas between San Francisco and Honolulu. Yesterday the Japanese Government also launched an attack against Malaya. Last night Japanese forces attacked Hong Kong. Last night Japanese forces attacked Guam. Last night Japanese forces attacked the Philippine Islands. Last night the Japanese attacked Wake Island. And this morning the Japanese attacked Midway Island. Japan has, therefore, undertaken a surprise offensive extending throughout the Pacific area. The facts of yesterday and today speak for themselves. The people of the United States have already formed their opinions and well understand the implications to the very life and safety of our Nation. As Commander in Chief of the Army and Navy I have directed that all measures be taken for our defense. But always will our whole Nation remember the character of the onslaught against us. No matter how long it may take us to overcome this premeditated invasion, the American people in their righteous might will win through to absolute victory. I believe that I interpret the will of the Congress and of the people when I assert that we will not only defend ourselves to the uttermost but will make it very certain that this form of treachery shall never again endanger us. Hostilities exist. There is no blinking at the fact that our people, our territory, and our interests are in grave danger. With confidence in our armed forces—with the unbounding determination of our people—we will gain the inevitable triumph, so help us God. I ask that the Congress declare that since the unprovoked and dastardly attack by Japan on Sunday, December 7, 1941, a state of war has existed between the United States and the Japanese Empire.

Source: Miller Center, University of Virginia. http://millercenterorg.pres ident/speeches/speech-3324. Accessed 12/3/2016.

A new basis of security and prosperity can be established for all

---

Franklin D. Roosevelt
A SECOND BILL OF RIGHTS

January 11, 1944

Introduction

America's Founding Fathers emphasized the importance of political rights, both by establishing a government with many external checks and balances and by incorporating specific restraints on governmental actions, most notably in the Bill of Rights. By contrast, many twentieth-century constitutions have focused on social and economic rights. In this fireside chat, which he delivered in place of his regular State of the Union Address, President Franklin D. Roosevelt proposed what he called a Second Bill of Rights, which would have formally acknowledged the expanded role that he had advocated for government in his New Deal programs. Although he did not directly address disaster relief, Roosevelt's emphasis on security would provide justification for expanded disaster relief. Although Roosevelt's proposals were not incorporated as amendments to the U.S. Constitution, modern disaster aid often includes grants and loans to provide housing, rebuild damaged infrastructure, provide unemployment insurance, and even compensation for those who have lost businesses and income.

This republic had its beginnings, and grew to its present strength, under the protection of certain inalienable political rights—among them the right of free speech, free press, free worship, trial by jury, freedom from unreasonable searches and seizures. They were our right to life and liberty.

We have come to a clear realization of the fact, however, that true individual freedom cannot exist without economic security and independence. "Necessitous men are not free men." People who are hungry, people who are out of a job are the stuff of which dictatorships are made.

In our day these economic truths have become accepted as self-evident. We have accepted, so to speak, a second Bill of Rights under which a new basis of security and prosperity can be established for all—regardless of station, or race or creed.

Among these are:

The right to a useful and remunerative job in the industries, or shops or farms or mines of the nation;

The right to earn enough to provide adequate food and clothing and recreation;

The right of farmers to raise and sell their products at a return which will give them and their families a decent living;

The right of every business man, large and small, to trade in an atmosphere of freedom from unfair competition and combination by monopolies at home or abroad;

The right of every family to a decent home;

The right to adequate protection from the economic fears of old age, and sickness, and accident and unemployment;

And finally, the right to a good education.

All of these rights spell security. And after this war is won we must be prepared to move forward, in the implementation of these rights, to new goals of human happiness and well-being.

Source: F.D.R. Fireside Chat 28: On the State of the Union. From Miller Center Website at the University of Virginia.

PART VI

The Cold War and Increasing Institutionalization of Disaster Responses

[A] plan of civil defense for the protection of life and property ... from attack

---

FEDERAL CIVIL DEFENSE ACT OF 1950

Introduction

Traditionally, one viewed national defense as defense against foreign enemies. This legislation, adopted after the Cold War had begun between the United States and the Soviet Union was designed to deal with catastrophes brought about by war, especially nuclear war. Much of the law involved identifying secure places that individuals could gather in case of such war. Throughout this period, there was often tension between governmental preparations for such man-made catastrophes and preparations for traditional natural catastrophes like hurricanes, tornadoes, and floods.

Title I -- Organization

Sec. 2. It is the policy and intent of Congress to provide a plan of civil defense for the protection of life and property in the United States from attack. It is further declared to be the policy and intent of Congress that this responsibility for civil defense shall be vested primarily in the several States and their political subdivisions. The Federal Government shall provide necessary coordination and guidance; shall be responsible for the operations of the Federal Civil Defense Administration as set forth in this Act; and shall provide necessary assistance as hereinafter authorized.

Sec. 101. (a) There is hereby established in the executive branch of the Government a Federal Civil Defense Administration (hereinafter referred to as the "Administration") at the head of which shall be a Federal Civil Defense Administrator appointed from civilian life by the President, by and with the consent of the Senate....

Title III – Emergency Authority

Sec. 301. The provisions of this title shall be operative only during the existence of a state of civil defense emergency (referred to hereinafter in this title as 'emergency'). The existe4nce of such emergency may be proclaimed by the President or by concurrent resolution of the Congress if the President in such proclamation, or the Congress in such resolution, finds that an attack upon the United States has occurred or is anticipated and that the national safety therefor requires an invocation of the provisions of this title ...

Source: For summary of this law, see Wilbur J. Cohen and Evelyn F. Boyer, *Bulletin* (April 1951), 11-16.

To promote and facilitate the civil defense of the United States in cooperation with the several states

---

Harry Truman
Executive Order 10186
ESTABLISHING THE FEDERAL CIVIL DEFENSE ADMINISTRATION IN THE OFFICE FOR EMERGENCY MANAGEMENT OF THE EXECUTIVE OFFICE OF THE PRESIDENT

December 1, 1950

Introduction

Just as the United States was emerging from its victory over fascism in World War II, it confronted the threat of communism as embodied in the regimes established in the Soviet Union and China. It was clear that many of the civil defense measures that the nation had utilized during World War II would need to continue in the new Cold War. Although the emphasis was clearly on the danger of foreign attacks, the Federal Civil Defense Administration, which this order established consistent with the Federal Civil Defense Act of 1950, would also be available in cases of natural disasters.

By virtue of the authority vested in me by the Constitution and the statutes, and in furtherance of the civil defense of the United States, it is ordered as follows:

1. There is hereby established the Federal Civil Defense Administration (hereinafter referred to as the Administration) in the Office for Emergency Management of the Executive Office of the President. At the head of the Administration shall be an Administrator who shall be appointed by the President with compensation at the rate of $17,500 a year, and who may appoint a Deputy Administrator with compensation at the rate of $16,000 a

year. The foregoing appointments shall be made without regard to the civil-service laws and the Classification Act of 1949.

2. The basic purpose of the Administration shall be to promote and facilitate the civil defense of the United States in cooperation with the several States. Subject to the direction and control of the President and within such amounts of funds as may be made available, and in accordance with law, the Administrator shall perform the following functions:

(a) Prepare comprehensive Federal plans and programs for the civil defense of the United States and coordinate them with the civil-defense activities of the States, of neighboring countries, and, with the consent of any such country, of any state, province, or similar political subdivision thereof.

(b) Conduct or arrange for the conduct of research to develop civil-defense measures and equipment and to effect the standardization thereof.

(c) Disseminate civil-defense information and exchange such information with foreign countries.

(d) Conduct or arrange for training programs for the instruction of State and local civil-defense leaders and specialists in the organization, operation, and techniques of civil defense.

(e) Assist and encourage any two States or groups of States or any one or more States and any neighboring state, province, or similar political subdivision of a foreign country, with the consent of such foreign country, in negotiating and entering into agreements or compacts for mutual aid across State lines, or into or out of the United States, to meet emergencies or disasters from enemy attacks which cannot be adequately met or controlled by the local forces: Provided, That all such agreements or compacts shall be subject to the consent of the Congress.

(f) Make appropriate provision for necessary civil-defense communications.

3. All departments and agencies of the Federal Government are authorized and directed to cooperate with the Administrator and, to the extent permitted by law, to furnish the Administrator such information and Assistance as he may require in the performance of his functions under this order. The Administrator shall, to the

extent practicable, utilize existing facilities and services of the departments and other agencies. The Administrator shall review the civil-defense activities of the departments and agencies and promote the coordination of these activities with one another and with the comprehensive Federal plans and programs prepared by the Administrator pursuant to this order. Insofar as the functions assigned hereunder relate to negotiations or the exchange of information with foreign countries or their political subdivisions, such functions shall be performed in cooperation with and subject to the approval of the Secretary of State.

4. To the extent necessary to carry out the provisions of this order, the Administrator is authorized (1) to employ civilian personnel for duty in the United States, including the District of Columbia, or elsewhere, subject to the civil-service laws, (2) to fix the compensation of such personnel in accordance with the Classification Act of 1949, and (3) to make provision for supplies, facilities, and services: Provided, That the rates of compensation for not more than twenty-two positions (1) may be fixed without regard to the Classification Act of 1949, (2) shall be not less than $11200 or more than $14,000 per annum, and (3) shall be fixed subject to the approval of the Civil Service Commission in those cases in which they are $13,000 or less per annum and subject to the approval of the President in those cases in which they are more than $13,000 per annum.

5. The Administrator is authorized to employ experts and consultants or organizations thereof, in accordance with and subject to the provisions of section 15 of the act of August 2, 1946 (5 U.S.C. 55a), at rates not in excess of $50 per diem. While away from their homes or regular places of business, on the business of the Administration, persons so employed may be paid actual transportation expenses and an allowance not to exceed $15 per diem in lieu of subsistence and other expenses.

6. Those activities with respect to civil defense heretofore performed by employees of the National Security Resources Board which are within the scope of this order shall hereafter be performed by the Administration; and the employees now primarily engaged in performing the said activities shall be transferred from the national

Security Resources Board to the Administration. The records of the National Security Resources Board relating to the said activities shall be made available to the Administrator pursuant to the provisions of Executive Order No. 9784 of September 25, 1946.

7. As used in this order, the terms "State" and "States" include the Territories and possessions of the United States and the District of Columbia.

8. Pending the appropriate of funds for the use of the Administration, its expenditures, including the compensation of personnel, shall be financed out of an allotment or allotments to be made by the President from the appropriation under the heading "Executive Office of the President-Emergencies (National Defense)" appearing in the Supplemental Appropriation Act, 1951 (Public Law 843, 81st Congress, approved September 27, 1950).

HARRY S. TRUMAN
THE WHITE HOUSE,
December 1, 1950

Source: Harry S. Truman: "Executive Order 10186—Establishing the Federal Civil Defense Administration in the Office for Emergency Management of the Executive Office of the President," December 1, 1950. Online by Gerhard Peters and John T. Woolley, *The American Presidency Project*. http://www.presidency.ucsb.edu/ws/?pid=78352.

The federal government shall provide necessary direction, coordination, and guidance

---

1958 AMENDMENTS TO CIVIL DEFENSE ACT OF 1950

Introduction

Throughout much of the Cold War, the focus of civil defense was on possible nuclear attacks. The national government saw itself as the central coordinator and encourager of, and as a major financial contributor to, state and local efforts.

Public Law 85-606

SEC. 2. Section 2 of the Act is amended by striking out said section and substituting the following therefor:

> "It is the sense of the Congress that the defense of the United States, in this thermonuclear age, can best be accomplished by enacting into law the measures set forth in this Act. It is the policy and intent of Congress to provide a system of civil defense for the protection of life and property in the United States from attack. It is further declared to be the policy and intent of the Congress that the responsibility for civil defense shall be vested jointly in the Federal Government and the several States and their political subdivisions. The Federal Government shall provide necessary direction, coordination, and guidance; shall be responsible for the operation of the Federal Civil Defense Administration as set forth in this Act; and shall provide necessary assistance as herein authorized."

Sec. 205. [Addition to previous law] To further assist in carrying out the purposes of this Act, the Administrator is authorized to make financial contributions to the States (including interstate civil defense authorities established pursuant to section 201 (g) of this Act) for necessary and essential State and local civil defense personnel and administrative expenses, on the basis of approved plans (which shall be consistent with the national plan for civil defense approved by the Administrator) for the civil defense of the States: Provided, That the financial contributions to the States for the purposes of this section shall not exceed one-half of the total cost of such necessary and essential State and local civil defense personnel and administrative expenses.

"[A] reasonable method of sharing the risk of flood losses is through a program of flood insurance which can complement and encourage preventive and protective measures"

NATIONAL FLOOD INSURANCE ACT OF 1968

FLOOD DISASTER PROTECTION ACT OF 1973

Introduction

More lives are lost, and property damaged, by flooding in America than through any other natural disaster. A vital component of the federal response to this issue has been the provision of national flood insurance, which would spread risks. Such legislation was, in turn, supposed to be tied to adopting land use policies and purchase insurance that would mitigate future disasters.

As with health care insurance, the financial solvency of the program depends in part in getting participation by individuals in areas with low flooding potential to help pay for rehabilitation for areas where flooding is more prevalent.

Sec. 4001. Congressional Findings and declaration of purpose

(a) Necessity and reasons for flood insurance program

The Congress finds that (1) from time to time flood disasters have created personal hardships and economic distress which have required unforeseen disaster relief measures and have placed an increasing burden on the Nation's resources; (2) despite the installation of preventive and protective works and the adoption of other public programs designed to reduce losses caused by flood damage, these methods have not been sufficient to protect adequately against growing exposure to future flood losses; (3) as a matter of national policy, a reasonable method of sharing the risk of flood losses is through a program of flood insurance which can complement and encourage preventive and protective measures; and (4) if such a program is initiated and carried out gradually, it can be expanded as knowledge is gained and experience is appraised, thus eventually making flood insurance coverage available on reasonable terms and conditions to persons who have need for such protection.

(b) Participation of Federal Government in flood insurance program carried out by private insurance industry.

The Congress also finds that (1) many factors have made it uneconomic for the private insurance industry alone to make flood insurance available to those in need of such protection on reasonable terms and conditions; but (2) a program of flood insurance with large-scale participation of the Federal Government and carried out to the maximum extent practicable by the private insurance industry is feasible and can be initiated.

(e) Land use adjustments by State and local governments; development of proposed future construction; assistance of lending and credit institutions; relation of Federal assistance to all flood-related programs; continuing studies.

It is the further purpose of this chapter to (1) encourage State and local governments to make appropriate land use adjustments to constrict the development of land which is exposed to flood damage and minimize damage caused by flood losses, (2) guide the development of proposed future construction, where practicable, away from locations which are threatened by flood hazards, (3) encourage lending and credit institutions, as a matter of national policy, to assist in furthering the objectives of the flood insurance program, (4) assure that any Federal assistance provided under the program will be related closely to all flood-related programs and activities of the Federal Government, and (5) authorize continuing studies of flood hazards in order to provide for a constant reappraisal of the flood insurance program and its effect on land use requirements.

Sec. 4002. Additional Congressional findings and declaration of purpose.

(c) The purpose of this Act, therefor, is to—

(1) Substantially increase the limits of coverage authorized under the national flood insurance program;

(2) Provide for the expeditious identification of, and the dissemination of information concerning, flood-prone areas;

(3) Require States or local communities, as a condition of future Federal financial assistance, to participate in the flood insurance program and to adopt adequate flood plain ordinances with effective enforcement provisions consistent with Federal standards to reduce or avoid future flood losses; and

(4) Require the purchase of flood insurance by property owners who are being assisted by Federal programs or by federally supervised, regulated, or insured agencies or institut8ions in the acquisition or improvement of land or facilities located or to be located in identified areas having special flood hazards.

Sec. 4012a. Flood insurance purchase and compliance requirements and escrow accounts.

(a) Amount and term of coverage

After the expiration of sixty days following December 31, 1973, no Federal officer or agency shall approve any financial assistance for acquisition or construction purposes for use in any area that has been identified by the Director as an area having special flood hazards and in which the sale of flood insurance has been made available under the National Flood Insurance Act unless the building or mobile home and any personal property to which such financial assistance relates is covered by flood insurance in an amount at least equal to its development or project cost (less estimated land cost) or to the maximum limit of coverage made available with respect to the particular type of property under the National Flood Insurance Act of 1968, whichever is less ...

Sec. 4022 State and local land use controls

(a) Requirement for participation in flood insurance program

(1) In general

After December 31, 1971, no new flood insurance coverage shall be provided under this chapter in any area (or subdivision thereof) unless an appropriate public body shall have adopted adequate land use and control measures (with effective enforcement provisions) which the Director finds are consistent with the comprehensive criteria for land management and use under section 4102 of this title.

Sec. 4028. Coastal Barrier Resources System

(a) No new flood insurance coverage may be provided under this chapter on or after October 1, 1983, for any new construction or substantial improvements of structures located on any coastal barrier within the coastal Barrier Resources System established by section 3503 of title 16. A federally insurance financial institution may make loans secured by structures which are not eligible for flood insurance by reason of this section.

Sec. 4104c. Mitigation assistance

(a) Authority

The Director shall carry out a program to provide financial assistance to States and communities, using amounts made available from the National Flood Mitigation Fund under section 4104d of this title, for planning and carrying out activities designed to reduce the risk of flood damage to structures covered under contracts for flood insurance under this chapter. Such financial assistance shall be made available to States and communities in the form of grants ...

Sec. 4106. Nonparticipation in flood insurance program

(a) Prohibition against Federal approval of financial assistance

No Federal officer or agency shall approve any financial assistance for acquisition or construction purposes on and after July 1, 1975, for use in any area that has been identified by the Director as an area having special flood hazards unless the community in which such area is situated is then participating in the national flood insurance program.

Source: All-hazard Authorities of the Federal Emergency Management Agency: The National Flood Insurance Act of 1968, as Amended, and The Flood Disaster Protection Act of 1973, as Amended, 42 U.S.C. 4001 *et. seq*. Office of the General Counsel, August, 1997, 91 pp.

Special measures designed to aid and accelerate the efforts to these affected States to reconstruct and rehabilitate the devastated areas

---

DISASTER RELIEF ACT OF 1969

Introduction

This law, initially prompted by floods and landslides in California over the winter of 1968-1969 and highlighted by Hurricane Camille, which struck Mississippi and Louisiana, sought to increase aid to private individuals in the aftermath of natural disasters. Note the discretion given to the president and the insistence on splitting the cost with state and local governments.

Public Law 91-79
AN ACT

To provide additional assistance for the reconstruction of areas damaged by major disasters.

Be it enacted by the Senate and House of Representatives of the United States of America in Congress assembled, That Congress hereby recognizes that a number of States have experiences extensive property loss and damage as a result of recent major disasters including, but not limited to, hurricanes, storms, floods, and high waters and wind-driven waters and that there is a need for special measures designed to aid and accelerate the efforts of these affected State to reconstruct and rehabilitate the devastated areas.

SEC. 2. The President is authorized to allocate funds hereafter appropriated to carry out this section to those States affected by a major disaster for the permanent repair and reconstruction of those permanent street, road, and highway facilities not on any of the Federal-aid systems which were destroyed or damaged as a result of such a major disaster. No funds shall be allocated under this section for repair or reconstruction of such a street, road, or highway facility unless the affected State agrees to pay not less than 50 per centum of all costs of such repair or reconstruction.

SEC. 3. Deals with timber contracts.

SEC. 6. Provided for disaster loans to small businesses.

SEC. 7. Provided for Emergency farm loans.

SEC 8. (a) The President is authorized to provide assistance to the States in developing comprehensive plans and practicable programs for assisting individuals suffering losses as the result of a major disaster...

(b) The President is authorized to make grants not to exceed $250,000 to any State, upon application therefor, in an amount not to exceed 50 per centum of the costs of developing the plans and programs referred to in subsection (a).

SEC. 9. The President shall, immediately upon his designation of an area as a major disaster area, appoint a Federal coordinating officer to operate under the Office of Emergency Preparedness in such area. Such officer shall be responsible for the coordination of all Federal disaster relief and assistance, shall establish such field offices as may be necessary for the rapid and efficient administration of Federal disaster relief programs, and shall otherwise assist local citizens and public officials in promptly obtaining assistance to which they are entitled.

SEC. 10 (a) The President is authorized to provide on a temporary basis, as prescribed in this section, dwelling accommodations for individuals and families displaced by a major disaster....

SEC. 11 (a) Whenever, as a result of a major disaster, the President determines that low-income households are unable to purchase adequate amounts of nutritious food, he is authorized, under such terms and conditions as he may prescribe, to distribute through the Secretary of Agriculture coupon Allotments to such households pursuant to provisions of the Food Stamp Act of 1964 and to make surplus commodities available pursuant to the provisions of section 3 of Public Law 875 of the Eighty-first Congress.

SEC. 14. The President, whenever he determines it to be in the public interest, and acting through the Director of the Office of

Emergency Preparedness, is authorized to make grants to any State or political subdivision thereof for the purpose or removing debris deposited on privately owned lands and on or in privately owned water as a result of a major disaster, and such States or political subdivision is authorized, upon application, to make payments to any person for reimbursement of expenses actually incurred by such person in the removal of such debris, but not to exceed the amount that such expenses exceed the salvage value of such debris.

[I]mproper development in flood-prone, storm surge, geological hazard, and erosion-prone areas

---

COASTAL ZONE MANAGEMENT ACT OF 1972

as amended through Pub. L. No. 10-9-58, the Energy Policy Act of 2005

Introduction

Much of the flood and hurricane damage that occurs in the U.S. takes place in coastal zones, which may be increasingly vulnerable as global warming raises sea levels. Because of their proximity to beaches, these are popular areas for development, and increasing numbers of people live there. This law, first adopted in 1972, and subsequently revised, demonstrates, however, that the government pursues multiple objectives, not all of which are always compatible.

The decision in *Lucas v. South Carolina Coastal Council* (1992), which is printed later in this book, also demonstrates that there are some constitutional limits on governmental regulation of property in this area.

16 U.S.C. Sec. 1451. Congressional findings (Section 3023)

The Congress finds that—

(a) There is a national interest in the effective management, beneficial use, protection, and development of the coastal zone.

(b) The coastal zone is rich in a variety of natural, commercial, recreational, ecological, industrial, and esthetic resources of immediate and potential value to the present and future well-being of the Nation.

(c) The increasing and competing demands upon the lands and waters of our coastal zone occasioned by population growth and economic development, including requirements for industry, commerce, residential development, recreation, extraction of mineral resources and fossil fuels, transportation and navigation, waste disposal, and harvesting of fish, shellfish, and other living marine resources, have resulted in the loss of living marine resources, wildlife, nutrient-rich areas, permanent and adverse changes to ecological systems, decreasing open space for public use, and shoreline erosion.

(d) The habitat areas of the coastal zone, and the fish, shellfish, other living resources, and wildlife therein, are ecol-ogically fragile and consequently extremely vulnerable to destruction by man's alterations

(e) Important ecological, cultural, historic, and esthetic values in the coastal zone which are essential to the well-being of all citizens are being irretrievably damaged or lost.

(f) New and expanding demands for food, energy, minerals, defense needs, recreation, waste disposal, transportation, and industrial activities in the Great Lakes, territorial sea, exclusive economic zone, and Outer Continental Shelf are placing stress on these areas and are creating the need for resolution of serious conflicts among important and competing uses and values in coastal and ocean waters;

(g) Special natural and scenic characteristics are being damaged by ill-planned development that threatens these values.

(h) In light of competing demands and the urgent need to protect and to give high priority to natural systems in the coastal zone, present state and local institutional arrange-ments for planning and regulating land and water uses in such areas are inadequate.

(i) The key to more effective protection and use of the land and water resources of the coastal zone is to encourage the states to exercise their full authority over the lands and waters in the coastal zone by assisting the states, in cooperation with Federal and local governments and other vitally affected interests, in developing land and water use programs for the coastal zone, including unified policies, criteria, standards, methods, and processes for dealing with land and water use decisions of more than local significance.

(j) The national objective of attaining a greater degree of energy self-sufficiency would be advanced by providing Federal financial assistance to meet state and local needs resulting from new or expanded energy activity in or affecting the coastal zone.

(k) Land uses in the coastal zone, and the uses of adjacent lands which drain into the coastal zone, may significantly affect the quality of coastal waters and habitats, and efforts to control coastal water pollution from land use activities must be improved.

(l) Because global warming may result in a substantial sea level rise with serious adverse affects in the coastal zone, coastal states must anticipate and plan for such an occurrence

(m) Because of their proximity to and reliance upon the ocean and its resources, the coastal states have substantial and significant interests in the protection, management, and development of the resources of the exclusive economic zone that can only be served by the active participation of coastal states in all Federal programs affecting such resources and, wherever appropriate, by the development of state ocean resource plans as part of their federally approved coastal zone management programs.

16 U.S.C. Sec. 1452, Congressional declaration of policy (Section 303)

The Congress finds and declares that it is the national policy—

(1) To preserve, protect, develop, and where possible, to restore or enhance, the resources of the Nation's coastal zone for this and succeeding generations;

(2) To encourage and assist the states to exercise effectively their responsibilities in the coastal zone through the development and implementation of management programs to achieve wise use of the land and water resources of the coastal zone, giving full consideration to ecological, cultural, historic, and esthetic values as well as the needs for compatible economic development, which programs should at least provide for—

(A) the protection of natural resources, including wetlands, floodplains, estuaries, beaches, dunes, barrier islands, coral reefs, and fish and wildlife and their habitat, within the coastal zone,

(B) the management of coastal development to minimize the loss of life and property caused by improper development in flood-prone, storm surge, geological hazard, and erosion-prone areas and in areas likely to be affected by or vulnerable to sea level rise, and subsidence, and saltwater intrusion, and by the destruction of natural protective features such as beaches, dunes, wetlands, and barrier islands,...

A declaration by the President that a major disaster exists

ROBERT T. STAFFORD DISASTER RELIEF AND EMERGENCY ASSISTANCE ACT

1988

Introduction

Almost as soon as natural disasters strike states, their governors typically apply to the president for a declaration of disaster relief, which opens up federal coffers. This policy began with the Stafford Act, which was originally enacted in 1974. It has been subsequently amended.

The law from which the following sections are excerpted is 81 pages, so the passages below only hit the highlights.

Sec. 101. Congressional Findings and Declarations (42 U.S.C. 5121)

(a) The Congress hereby finds and declares that –

(1) Because disasters often cause loss of life, human suffering, loss of income, and property loss and damage; and

(2) Because disasters often disrupt the normal functioning of governments and communities, and adversely affect individuals and families with great severity; special measures, designed to assist the efforts of the affected States in expediting the rendering of aid, assistance, and emergency services, and the reconstruction and rehabilitation of devastated areas, are necessary.

(b) It is the intent of the Congress, by this Act, to provide an orderly and continuing means of assistance by the Federal Government to State and local governments in carrying out their responsibilities to alleviate the suffering and damage which result from such disasters by –

(1) Revising and broadening the scope of existing disaster relief programs;

(2) Encouraging the development of comprehensive disaster preparedness and assistance plans, programs, capabilities, and organizations by the State and by local governments;

(3) Achieving greater coordination and responsiveness of disaster preparedness and relief programs;

(4) Encouraging individuals, States, and local governments to protect themselves by obtaining insurance coverage to supplement or replace governmental assistance;

(5) Encouraging hazard mitigation measures to reduce losses from disasters, including development of land use and construction regulations; and

(6) Providing Federal assistance program for both public and private losses sustained in disasters.

Sec. 201 – Federal and State Disaster Preparedness Programs (42 U.S.C. 5131)

(a) UTILIZATION OF SERVICES OF OTHER AGENCIES –The President is authorized to establish a program of disaster preparedness that utilizes services of all appropriate agencies and includes –

(1) preparation of disaster preparedness plans for mitigation, warming, emergency operations, rehabilitation, and recovery;

(2) training and exercises;

(3) postdisaster critiques and evaluations;

(4) annual review of programs;

(5) coordination of Federal, State, and Local preparedness programs;

(6) application of science and technology;

(7) research

Sec. 401. Procedure for Declaration (42 U.S.C. 5170)

(a) IN GENERAL – All requests for a declaration by the President that a major disaster exists shall be made by the Governor of the affected State. Such a request shall be based on a finding that the disaster is of such severity and magnitude that effective response is beyond the capabilities of the State and the affected local governments and that Federal assistance is necessary. As part of such request, and as a prerequisite to major disaster assistance under this Act, the Governor shall take appropriate response action under State law and direct execution of the State's emergency plan. The Governor shall furnish information on the nature and amount of State and local resources which have been or will be committed to alleviating the results of the disaster, and shall certify that, for the current disaster, State and local government obligations and expenditures (of which State commitments must be a significant proportion) will comply with all applicable cost-sharing requirements of this Act. Based on the request of a Governor under this section, the President may declare under this Act that a major disaster or emergency exists.

Sec. 404. Hazard Mitigation
(42 U.S.C. 5170c)

(a) IN GENERAL – The President may contribute up to 75 percent of the cost of hazard mitigation measures which the President has determined are cost-effective and which substantially reduce the risk of future damage, hardship, loss, or suffering in any area affected by a major disaster. ...

Sec. 408. Federal Assistance to Individuals and Households
(42 U.S.C. 5174)

(h) MAXIMUM AMOUNT OF ASSISTANCE –

(1) In GENERAL – No individual or household shall receive financial assistance greater than $25,000 under this section with respect to a single major disaster.

Source: The Stafford Act, Robert T. Stafford Disaster Relief and Emergency Assistance Act, *as Amended*. April, 2013. Department of Homeland Security. FEMA.

Fire is an undue burden... and... constitutes a public health and safety problem of great dimensions

---

FEDERAL FIRE PREVENTION AND CONTROL ACT OF 1974

(Public Law 93-498. As amended)

Introduction

Although historic fires (such as the Chicago fire of 1871) have wiped out entire cities, or major portions thereof, or entire buildings (as in the March 25, 1911 Triangle Shirtwaist Company factory in New York City that killed 145 workers and led to stricter building codes) as well as swaths of major forests, because most are much smaller phenomena, most do not capture the public attention in the same fashion as floods or hurricanes. Fire safety nonetheless is a major cause of damage each year. This law, a small excerpt from almost 50 pages of text, demonstrates that the national government is also concerned with the issue.

Sec. 2. The Congress finds that –

(1) The National Commission on Fire Prevention and Control, established pursuant to Public Law 90-259, has made an exhaustive and comprehensive examination of the Nation's fire problem, has made detailed findings as to the extent of this problem in terms of human suffering and loss of life and property, and has made ninety thoughtful recommendations.

(2) The United States today has the highest per capita rate of death and property loss from fire of all the major industrialized nations in the world.

(3) Fire is an undue burden affecting all Americans, and fire also constitutes a public health and safety problem of great dimensions. Fire kills 12,000 and scars and injures 300,000 Americans each year, including 50,000 individuals who require

extended hospitalization. Almost $3 billion worth of property is destroyed annually by fire, and the total economic cost of destructive fire in the United States is estimated conservatively to be $11,000,000,000 per year. Firefighting is the Nation's most hazardous profession.

(4) Such losses of life and property from fire are unacceptable to the Congress.

(5) While fire prevention and control is and should remain a State and local responsibility, the federal Government must help if a significant reduction in fire losses is to be achieved.

(6) The fire service and the civil defense program in each locality would both benefit from closer cooperation.

(7) The Nation's fire problem is exacerbated by (A) the indifference with which some Americans confront the subject; (B) the Nation's failure to undertake enough research and development into fire and fire-related problems; (C) the scarcity of reliable data and information; (D) the fact that designers and purchasers of buildings and products generally give insufficient attention to fire safety; € the fact that many communities lack adequate building and fire prevention codes; and (F) the fact that local fire departments spend about 95 cents of every dollar appropriated to the fire services on efforts to extinguish fires and only about 5 cents on fire prevention.

(8) There is a need for improved professional training and education oriented toward improve the effectiveness of the fire services, including an increased emphasis on preventing fires and on reducing injuries to firefighters.

(9) A national system for the collection, analysis, and dissemination of fire data is needed to help local fire services establish research and action priorities.

(10) The number of specialized medical centers which are properly equipped and staffed for the treatment of burns and the rehabilitation of victims of fires is inadequate.

(11) The unacceptably high rates of death, injury, and property loss from fire can be reduced if the federal Government establishes a coordinated program to support and reinforce the fire prevention and control activities of State and local governments.

Sec. 2. It is declared to be the purpose of Congress in this Act to—

(1) reduce the Nation's losses caused by fire through better fire prevention and control;

(2) supplement existing programs of research, training, and education, and to encourage new and improved programs and activities by State and local governments;

(3) establish the United States Fire Administration and the Fire Research Center within the Department of Commerce; and

(4) establish an intensified program of research into the treatment of burn and smoke injuries and the rehabilitation of victims of fires within the National Institutes of Health.

Special measures are necessary to assist the efforts of the States

---

FEDERAL DISASTER RELIEF ACT OF 1974

Introduction

The Disaster Relief Act of 1974 reflected both the national government's desire to respond more comprehensively to disasters and to encourage states and localities to take preventative measures to prepare for them. Schneider (2011, p. 20) observes that the laws "stressed a 'multihazard' approach to disasters," which she describes as equipping governments to "be capable of handling *all* kinds of hazards—both natural and manmade."

Public Law No: 93-288 (05/22/1974)

Disaster Relief Act - Title I: Findings, Declarations, and Definitions= - Declares the finding of Congress that special measures are necessary to assist the efforts of the States in expediting the rendering of aid and emergency services and reconstruction and rehabilitation of devastated areas.

Defines the terms used in this Act.

Title II: Disaster Preparedness Assistance - Authorizes the President to establish a program of disaster preparedness that utilizes services of all appropriate agencies.

Directs the President to provide technical assistance to the States in developing comprehensive plans and practicable programs for preparation against disasters.

Provides that, upon application by a State, the President is authorized to make grants, not to exceed in the aggregate to such State $250,000, for the development of plans, programs, and capabilities for disaster preparedness and prevention. Authorizes the President to make grants not to exceed 50 percent of the cost of improving, maintaining, and updating State disaster assistance plans, with a grant limit of $25,000 per year per State.

Directs the President to insure that all appropriate Federal agencies are prepared to issue warnings of disasters to State and local officials.

Title III: Disaster Assistance Administration - Provides that all requests for a determination by the President that an emergency or a major disaster exists shall be made by the Governor of the affected State. Provides that, based upon such request, the President may declare that an emergency or a major disaster exists.

Authorizes the President to direct any Federal agency, with or without reimbursement, to utilize its available personnel, equipment, supplies, facilities, and other resources in support of State and local disaster assistance efforts.

Directs the President to appoint a Federal coordinating officer to operate in the affected area of a declared major disaster. Sets forth the duties of such officer.

Directs the President to form emergency support teams of Federal personnel to be deployed in an area affected by a major disaster or emergency to assist the Federal coordinating officer.

Sets forth authorized forms of assistance to be given in a major disaster or emergency area.

Title IV: Federal Disaster Assistance Programs - Authorizes the President to authorize any Federal agency to repair or reconstruct any facility owned by the United States and under the jurisdiction of such agency, which is damaged or destroyed by any major disaster.

Authorizes the President to make contributions to State or local governments to help repair or reconstruct public facilities belonging to such State or local governments, which were damaged or destroyed by a major disaster.

Authorizes the President to make grants to help repair or reconstruct private nonprofit educational, utility, emergency, medical, and custodial care facilities which were damaged by a major disaster.

Authorizes the President to provide, either by purchase or lease, temporary housing for those who require it as a result of a major disaster. Authorizes the President to provide assistance on a temporary basis in the form of mortgage or rental payments to or on behalf of those who, as a result of financial hardship caused by a major disaster, have received written notice of dispossession or eviction.

Authorizes the President to provide to any individual unemployed as a result of a major disaster such benefit assistance as he deems appropriate while such individual is unemployed.

Authorizes the President to make a grant to a State for the purpose of such State making grants to meet disaster - related necessary expenses or serious needs of individuals or families adversely affected by a major disaster in those cases where such individuals or families are unable to meet such expenses or needs through assistance under other provisions of this Act, or from other means.

Authorizes the President to make loans to any local government which may suffer a substantial loss of tax and other revenues as a result of a major disaster, and has demonstrated a need for financial assistance in order to perform its governmental functions. Sets forth the procedures for administration of such loans.

Authorizes additional relief programs for major disaster areas, including distribution of food coupons and food commodities, relocation assistance, legal services, crisis counseling assistance and training, emergency communications, emergency public transportation, and fire suppression grants.

Provides that where an existing timber sale contract between the Secretary of Agriculture or the Secretary of the Interior and a timber purchaser does not provide relief from major physical change not due to negligence of the purchaser prior to approval of construction of any section of specified road or of any other specified development facility and, as a result of a major disaster, a major physical change results in additional construction work in connection with such road or facility by such purchaser, the increased construction cost shall be borne by the United States.

Title V: Economic Recovery for Disaster Areas - Revises the Public Works and Economic Development Act of 1965 by providing that in the case of any area affected by a major disaster, the State Governor may request the President for assistance for economic recovery under this Act. Requires the Governor to designate a Recovery Planning Council for such area. Sets forth the membership composition of the Council. Provides that the purpose of the Council is to provide cooperative planning for development, restoration of employment base, and continued coordination of Federal-aid programs for long-range restoration and rehabilitation of normal commercial, industrial, and other economic activities. Sets forth administrative procedures of the Council.

Authorizes the President to provide funds to any Recovery Planning Council for the implementation of a recovery investment plan by public bodies. Sets forth regulations with respect to the use and administration of such funds.

Authorizes the appropriation of not in excess of $250,000,000 to carry out the provisions of this title.

Title VI: Miscellaneous - Authorizes the President to prescribe such rules and regulations as may be necessary and proper to carry out any of the provisions of this Act.

Makes various technical and conforming amendments.

Authorizes appropriations to the President of such sums as may be necessary to carry out this Act through the close of June 30, 1977.

Summary authored by Congressional Research Service (CRS).

Source: http://www.congress.gov/bill/93rd-congress/senate-bill/3062 Congress.gov.

All functions... are transferred or reassigned to the Director of the Federal Emergency Management Agency

---

Jimmy Carter
EXECUTIVE ORDER 12148
FEDERAL EMERGENCY MANAGEMENT

July 20, 1979

Introduction

From time to time, presidents seek to reorganize the administration of government. In this case, President Jimmy Carter, who served from 1977-1981, sought to combine functions that laws had previously dispersed throughout the national government into a single agency, the Federal Emergency Management Agency (FEMA), to be headed by a single Director. The opening paragraph of the order gives an idea of the way that disaster responses had previously been parceled out among diverse federal agencies. After the terrorist attacks of September 11, 2001, President George W. Bush incorporated FEMA within the Department of Homeland Security.

By the authority vested in me as President by the Constitution and laws of the United States of America, including the Federal Civil Defense Act of 1950, as amended (50 U.S.C. App. 2251 et seq.), the Disaster Relief Act of 1970, as amended (42 U.S.C. Chapter 58 note), the Disaster Relief Act of 1974 (88Stat. 143; 42 U.S.C. 5121 et seq.), the Earthquake Hazards Reduction Act of 1977 (42 U.S.C. 77-1 et seq.), Section 4 of Public Law 92-385 (86 Stat.556), Section 43 of the Act of august 10, 1856, as amended (50 U.S.C. App. 2285), the National Security Act of 1947, as amended, the Defense Production Act of 1050, as amended (50 U.S.C. App. 2961 et seq.), Reorganization Plan No. 1 of 1958, Reorganization Plan No. 1 of 1973, the Strategic and Critical Materials Stock Piling Act, as amended (50 U.S.C. 98 et seq.), Section 202 of the Budget and Accounting Procedures Act of 1940 (31 U.S.C. 585c), and Section 302 of Title 3 of the United States Code, and in order to transfer emergency functions to the federal Emergency Management Agency, it is hereby ordered as follows:

1-1-. Transfer or Reassignment of Existing Functions

1-10-. All functions vested in the President that have been delegated or assigned to the Defense Ci8vil Preparedness Agency, Department of defense, are transferred or reassigned to the Director of the Federal Emergency Management Agency.

1-102. All functions vested in the President that have been delegated or assigned to the Federal Disaster Assistance Administration, Department of Housing and Urban Development, are transferred or reassigned to the Director of the Federal Emergency Management Agency, including any of those functions redelegated or reassigned to the Department of Commerce with respect to assistance to communities in the development of readiness plans for severe weather-related emergencies.

1-103. All functions vested in the President that have been delegates or assigned to the Federal Preparedness Agency, General Services Administration, are transferred or reassigned to the Director of the Federal Emergency Management Agency.

1-1-4. All functions vested in the President by the Earthquake Hazards Reduction Act of 1944 (42 U.S.C.770-1 et seq.), including those functions performed by the Office of Science and Technology Policy, are delegates, transferred, or reassigned to the Director of the Federal Emergency Management Agency.

2-201. In executing the functions under this Order, the Director shall develop policies which provide that all civil defense and civil emergency functions, resources, and systems of Executive agencies are:

(a) founded on the use of existing organizations, resources, and systems to the maximum extent practicable.

(b) Integrated effectively with organizations, resources, and programs of State and local governments, the private sector and volunteer organizations; and

(c) Developed, tested and utilized to prepare for, mitigate, respond to and recover from the effects on the population of all forms of emergencies.

Source: The American Presidency Project. http://www.presidency.ucsb.edu/ws/?pid=32625. Accessed 11/25/2016.

[T]he national contingency plan for the removal of oil and hazardous substances

---

COMPREHENSIVE ENVIRONMENTAL RESPONSE COMPENSATION, AND LIABILITY ACT

42 U.S.C. Sec 9601 ff. (1980)

Introduction

Love Canal, a site near Niagara Falls, New York, is but one site of industrial pollution that threatened an entire neighborhood. It is a reminder that while it rarely gets the attention of a tornado, hurricane, or flood, long-term exposure to toxins can have disastrous consequences.

In 1980, Congress adopted a Comprehensive Environmental Response, Compensation, and Liability Act (CERCLA). It was commonly known as the Superfund, because it collected taxes from chemical and petroleum companies to clean up toxic waste sites as identified by the Environmental Protection Agency (EPA). This law has been further amended and refined by the Superfund Amendments and Reauthorization Act (SARA), which was adopted on October 17, 1986.

Sec 9601

The term "act of God"means an unanticipated grave natural disaster or other natural phenomenon of an exceptional, inevitable, and irresistible character, the effects of which could not have been prevented or avoided by the exercise of due care or foresight.

Sec. 9604

(a) Removal and other remedial action by President; applicability of national contingency plan; response by potentially responsible parties; public health threats; limitations on response; exception.

(1) Whenever (A) any hazardous substance is released or there is a substantial threat of such release into the environment, or (B) there is a release or substantial threat of release into the environment of any pollutant or contaminant which may present an imminent and substantial danger to the public health or welfare, the President is authorized to act, consistent with the national contingency plan, to remove or arrange for the removal of, and provide for remedial action relating to such hazardous substance, pollutant, or contaminant at any time (including its removal from any contaminated natural resource), or take any other response measure consistent with the national contingency plan which the President deems necessary to protect the public health or welfare or the environment....

Sec. 9605

National contingency plan

(a) Revision and republication

Within one hundred and eighty days after December 11, 1980, the President shall, after notice and opportunity for public comments, revise and republish the national contingency plan for the removal of oil and hazardous substances... such revision shall include a section of the plan to be known as the national hazardous substance response plan which shall establish procedures and standards for responding to releases of hazardous substances, pollutants, and contaminants, which shall include at a minimum:

(1) Methods for discovering and investigating facilities at which hazardous substances have been disposed of or otherwise come to be located;

(2) Methods for evaluation, including analyses of relative cost, and remedying any releases or threats of releases from facilities which pose substantial dander to the public health or the environment;

(3) Methods and criteria for determining the appropriate extent of removal, remedy, and other measures authorized by this chapter;

(4) Appropriate roles and responsibilities for the Federal, State, and local governments and for interstate and nongovernmental entities in effectuating the plan;

(5) Provision for identification, procurement, maintenance, and storage of response equipment and supplies;

(6) A method for and assignment of responsibility for reporting the existing of such facilities which may be located on federally owned or controlled properties and any releases of hazardous substances from such facilities;

(7) Means of assuring that remedial action measures are cost-effective over the period of potential exposure to the hazardous substances or contaminated materials;

(8) (A) criteria for determining priorities among releases or threatened releases throughout the United States for the purpose of taking remedial action and, to the extent practicable taking into account the potential urgency of such action, for the purpose of taking removal action....

[F]ostering international co-operation in the field of natural disaster reduction

---

U.N. DECLARATION DESIGNATING THE 1990S AS THE INTERNATIONAL DECADE FOR NATURAL DISASTER REDUCTION

December 11, 1987

Introduction

Although this book largely concentrates on disasters that occur within the United States, such events are a world-wide problem. Indeed, disasters typically fall far harder on underdeveloped economies without adequate resources and infrastructures than it does upon the United States and the rest of the developed world. This resolution from the General Assembly of the United Nations designating the 1990s as a decade for natural disaster reduction thus serves to put U.S. problems within a larger world-wide context.

...Considering that natural disasters, such as earthquakes, windstorms (cyclones, hurricanes, tornadoes, typhoons), tsunamis, floods, landslides, volcanic eruptions, wildfires and other calamities of natural origin, have claimed about 3 million lives worldwide in the past two decades, adversely affected the lives of at least 800 million more people and resulted in immediate damages exceeding $23 billion,

Considering also that, among disasters of natural origin, drought and Desertification are resulting in enormous damage, particularly in Africa, where the recent drought threatened the lives of more than 20 million people and uprooted millions of others,

Recognizing that the effects of such disasters may damage very severely the fragile economic infrastructure of developing countries, especially the least developed, land-locked and island developing countries, and thus hamper their development process,

Recalling the report of the Secretary-General on the work of the Organization, particularly the section concerning natural disasters and the merits of proposals that have been made to stimulate international study, planning and preparations on this subject over the next decade under the auspices of the United Nations,

Also taking note with appreciation of the report of the Secretary-General concerning the existing mechanisms and arrangements within the United Nations system for disaster and emergency assistance and co-ordination,

Recognizing the responsibility of the United Nations system for promoting international co-operation in the study of natural disasters of geophysical origin and in the development of techniques to mitigate risks arising therefrom, as well as for co-ordinating disaster relief, preparedness and prevention, including prediction and early warning,

Convinced that concerted international action for the reduction of natural disasters over the course of the 1990s would give genuine impetus to a series of concrete measures at the national, regional and international levels,

Recognizing that the primary responsibility for defining the general goals and directions of efforts undertaken in the framework of an international decade for natural disaster reduction and for implementing the measures that would result from the activities of the decade lies with the Governments of the countries concerned,

Considering that the concept of a global programme for natural disaster reduction is predicated on collaborative efforts among culturally and economically diverse nations, together with relevant organizations of the United Nations system and concerned national and international non-governmental organizations, including scientific and technological institutions,

1. Recognizes the importance of reducing the impact of natural disasters for all people, and in particular for developing countries;

2. Recognizes further that scientific and technical understanding of the causes and impact of natural disasters and of ways to reduce both human and property losses has progressed to such an

extent that a concerted effort to assemble, disseminate and apply this knowledge through national, regional and world-wide programmes could have very positive effects in this regard, particularly for developing countries;

3. Decides to designate the 1990s as a decade in which the international community, under the auspices of the United Nations, will pay special attention to fostering international co-operation in the field of natural disaster reduction, and to take a decision at its forty-third session on the content and modalities of United Nations participation therein after having considered the report of the Secretary-General referred to in paragraph 9 of the present resolution;

4. Decides that the objective of this decade is to reduce through concerted international actions, especially in developing countries, loss of life, property damage and social and economic disruption caused by natural disasters, such as earthquakes, windstorms (cyclones, hurricanes, tornadoes, typhoons), tsunamis, floods, landslides, volcanic eruptions, wildfires and other calamities of natural origin, such as grasshopper and locust infestations, and that its goals are:

(a) To improve the capacity of each country to mitigate the effects of natural disasters expeditiously and effectively, paying special attention to assisting developing countries in the establishment, when needed, of early warning systems;

(b) To devise appropriate guidelines and strategies for applying existing knowledge, taking into account the cultural and economic diversity among nations;

(c) To foster scientific and engineering endeavours aimed at closing critical gaps in knowledge in order to reduce loss of life and property;

(d) To disseminate existing and new information related to measures for the assessment, prediction, prevention and mitigation of natural disasters;

(e) To develop measures for the assessment, prediction, prevention and mitigation of natural disasters through programmes of technical assistance and technology transfer, demonstration projects, and education and training, tailored to specific hazards and locations, and to evaluate the effectiveness of those programmes;

5. Requests the Secretary-General, in co-operation with the appropriate organizations of the United Nations system and relevant scientific, technical, academic and other non-governmental organizations, to develop an appropriate framework for attaining the objective and goals referred to in paragraphs 3 and 4 above and to submit a report thereon to the General Assembly at its forty-fourth session through the Economic and Social Council;

6. Recommends that, if necessary, extrabudgetary resources be provided for the preparation of the above-mentioned report and considers that, for this purpose, voluntary contributions from countries, international organizations and other organizations are highly desirable;

7. Calls upon all Governments to participate during the decade for concerted international action for the reduction of natural disasters and, as appropriate, to establish national committees, in co-operation with the relevant scientific and technological communities, with a view to surveying available mechanisms and facilities for the reduction of natural hazards, assessing the particular requirements of their respective countries or regions in order to add to, improve or update existing mechanisms and facilities and develop a strategy to attain the desired goals;

8. Further calls upon Governments to keep the Secretary-General informed of their countries' plans and of assistance that can be provided so that the United Nations may become an international centre for the exchange of information, the storing of documents and the co-ordination of international efforts concerning the activities in support of the objective and goals referred to in paragraphs 3 and 4 above, thus enabling each Member State to benefit from the experience of other countries;

9. Requests the Secretary-General to report to the General Assembly at its forty-third session on progress made in the preparations outlined above with particular emphasis on defining the catalytic and facilitating role envisaged for the United Nations system.

Source: http://www.un.org/documents/ga/res/42/a42r169.htm.

[W]here regulation denies all economically beneficial or productive use of land

---

LUCAS V. SOUTH CAROLINA COASTAL COUNCIL

505 U.S. 1003; 112 S. Ct. 2886; 120 L. Ed. 2d 798 (1992)

Introduction

The government could lessen the occasions where it provided humanitarian disaster assistance if it could prevent certain risky behaviors, including building houses in flood zones. In this case, however, the U.S. Supreme Court decided that laws that seek to prevent building on previously unregulated lands could sometimes amount to a Fifth Amendment Taking, which requires governmental compensation.

The decision, written by Justice Antonin Scalia, reviews relevant precedents. Concurring and dissenting opinions have been omitted.

Justice Scalia delivered the opinion of the Court.

In 1986, petitioner David H. Lucas paid $975,000 for two residential lots on the Isle of Palms in Charleston County, South Carolina, on which he intended to build single-family homes. In 1988, however, the South Carolina Legislature enacted the Beachfront Management Act, S. C. Code Ann. § 48-39-250 et seq. (Supp. 1990), which had the direct effect of barring petitioner from erecting any permanent habitable structures on his two parcels. See § 48-39-290(A). A state trial court found that this prohibition rendered Lucas's parcels "valueless." App. to Pet. for Cert. 37. This case requires us to decide whether the Act's dramatic effect on the economic value of Lucas's lots accomplished a taking of private property under the Fifth and Fourteenth Amendments requiring the payment of "just compensation." U.S. Const., Amdt. 5.

I

A

South Carolina's expressed interest in intensively managing development activities in the so-called "coastal zone" dates from 1977 when, in the aftermath of Congress's passage of the federal Coastal Zone Management Act of 1972, 86 Stat. 1280, as amended, 16 U.S.C. § 1451 et seq., the legislature enacted a Coastal Zone Management Act of its own.... In its original form, the South Carolina Act required owners of coastal zone land that qualified as a "critical area" (defined in the legislation to include beaches and immediately adjacent sand dunes, § 48-39-10(J)) to obtain a permit from the newly created South Carolina Coastal Council (Council) (respondent here) prior to committing the land to a "use other than the use the critical area was devoted to on [September 28, 1977]." § 48-39-130(A).

In the late 1970's, Lucas and others began extensive residential development of the Isle of Palms, a barrier island situated eastward of the city of Charleston. Toward the close of the development cycle for one residential subdivision known as "Beachwood East," Lucas in 1986 purchased the two lots at issue in this litigation for his own account. No portion of the lots, which were located approximately 300 feet from the beach, qualified as a "critical area" under the 1977 Act; accordingly, at the time Lucas acquired these parcels, he was not legally obliged to obtain a permit from the Council in advance of any development activity. His intention with respect to the lots was to do what the owners of the immediately adjacent parcels had already done: erect single-family residences. He commissioned architectural drawings for this purpose.

The Beachfront Management Act brought Lucas's plans to an abrupt end. Under that 1988 legislation, the Council was directed to establish a "baseline" connecting the landwardmost "points of erosion... during the past forty years" in the region of the Isle of Palms that includes Lucas's lots.... In action not challenged here, the Council fixed this baseline landward of Lucas's parcels. That was significant, for under the Act construction of occupyable improvements was flatly prohibited seaward of a line drawn 20 feet landward of, and parallel to, the baseline. § 48-39-290(A). The Act provided no exceptions.

B

Lucas promptly filed suit in the South Carolina Court of Common Pleas, contending that the Beachfront Management Act's construction bar effected a taking of his property without just compensation. Lucas did not take issue with the validity of the Act as a lawful exercise of South Carolina's police power, but contended that the Act's complete extinguishment of his property's value entitled him to compensation regardless of whether the legislature had acted in furtherance of legitimate police power objectives. Following a bench trial, the court agreed. Among its factual determinations was the finding that "at the time Lucas purchased the two lots, both were zoned for single-family residential construction and... there were no restrictions imposed upon such use of the property by either the State of South Carolina, the County of Charleston, or the Town of the Isle of Palms." App. to Pet. for Cert. 36. The trial court further found that the Beachfront Management Act decreed a permanent ban on construction insofar as Lucas's lots were concerned, and that this prohibition "deprived Lucas of any reasonable economic use of the lots,... eliminated the unrestricted right of use, and rendered them valueless." Id., at 37. The court thus concluded that Lucas's properties had been "taken" by operation of the Act, and it ordered respondent to pay "just compensation" in the amount of $1,232,387.50. The Supreme Court of South Carolina reversed. It found dispositive what it described as Lucas's concession "that the Beachfront Management Act [was] properly and validly designed to preserve... South Carolina's beaches." 304 S.C. 376, 379, 404 S.E.2d 895, 896 (1991). Failing an attack on the validity of the statute as such, the court believed itself bound to accept the "uncontested... findings" of the South Carolina Legislature that new construction in the coastal zone -- such as petitioner intended -- threatened this public resource. Id., at 383, 404 S.E.2d at 898. The court ruled that when a regulation respecting the use of property is designed "to prevent serious public harm," id., at 383, 404 S.E.2d at 899 (citing, inter alia, *Mugler v. Kansas,* 123 U.S. 623, 31 L. Ed. 205, 8 S. Ct. 273 (1887)), no compensation is owing under the Takings Clause regardless of the regulation's effect on the property's value.

Two justices dissented. They acknowledged that our Mugler line of cases recognizes governmental power to prohibit "noxious" uses of property -- i.e., uses of property akin to "public nuisances" -- without having to pay compensation. But they would not have characterized the Beachfront Management Act's "primary purpose [as] the prevention of a nuisance." 304 S.C. at 395, 404 S.E.2d at 906 (Harwell, J., dissenting). To the dissenters, the chief purposes of the legislation, among them the promotion of tourism and the creation of a "habitat for indigenous flora and fauna," could not fairly be compared to nuisance abatement. Id., at 396, 404 S.E.2d at 906. As a consequence, they would have affirmed the trial court's conclusion that the Act's obliteration of the value of petitioner's lots accomplished a taking.

We granted certiorari. 502 U.S. 966 (1991).

II

As a threshold matter, we must briefly address the Council's suggestion that this case is inappropriate for plenary review. After briefing and argument before the South Carolina Supreme Court, but prior to issuance of that court's opinion, the Beachfront Management Act was amended to authorize the Council, in certain circumstances, to issue "special permits" for the construction or reconstruction of habitable structures seaward of the baseline. See S. C. Code Ann. § 48-39-290(D)(1) (Supp. 1991). According to the Council, this amendment renders Lucas's claim of a permanent deprivation unripe, as Lucas may yet be able to secure permission to build on his property. "[The Court's] cases," we are reminded, "uniformly reflect an insistence on knowing the nature and extent of permitted development before adjudicating the constitutionality of the regulations that purport to limit it." MacDonald, *Sommer & Frates v. Yolo County,* 477 U.S. 340, 351, 91 L. Ed. 2d 285, 106 S. Ct. 2561 (1986).. Because petitioner "has not yet obtained a final decision regarding how [he] will be allowed to develop [his] property," ... the Council argues that he is not yet entitled to definitive adjudication of his takings claim in this Court.

We think these considerations would preclude review had the South Carolina Supreme Court rested its judgment on ripeness grounds, as it was (essentially) invited to do by the Council. See Brief for Respondent 9, The South Carolina Supreme Court shrugged off the possibility of further administrative and trial proceedings, however, preferring to dispose of Lucas's takings claim on the merits..... This unusual disposition does not preclude Lucas from applying for a permit under the 1990 amendment for future construction, and challenging, on takings grounds, any denial. But it does preclude, both practically and legally, any takings claim with respect to Lucas's past deprivation, i.e., for his having been denied construction rights during the period before the 1990 amendment.... Without even so much as commenting upon the consequences of the South Carolina Supreme Court's judgment in this respect, the Council insists that permitting Lucas to press his claim of a past deprivation on this appeal would be improper, since "the issues of whether and to what extent [Lucas] has incurred a temporary taking... have simply never been addressed." Brief for Respondent 11. Yet Lucas had no reason to proceed on a "temporary taking" theory at trial, or even to seek remand for that purpose prior to submission of the case to the South Carolina Supreme Court, since as the Act then read, the taking was unconditional and permanent. Moreover, given the breadth of the South Carolina Supreme Court's holding and judgment, Lucas would plainly be unable (absent our intervention now) to obtain further state-court adjudication with respect to the 1988-1990 period.

In these circumstances, we think it would not accord with sound process to insist that Lucas pursue the late-created "special permit" procedure before his takings claim can be considered ripe. Lucas has properly alleged Article III injury in fact in this case, with respect to both the pre-1990 and post-1990 constraints placed on the use of his parcels by the Beachfront Management Act. 3 That there is a discretionary "special permit" procedure by which he may regain -- for the future, at least -- beneficial use of his land goes only to the prudential "ripeness" of Lucas's challenge, and for the reasons discussed we do not think it prudent to apply that prudential requirement here. See *Esposito v. South Carolina Coastal Council*, 939 F.2d 165, 168 (CA4 1991), cert. denied, post, 505 U.S. 1219.4 We leave for decision on remand, of course, the questions left unaddressed by the South Carolina Supreme Court as a consequence of its categorical disposition.

III

A

Prior to Justice Holmes's exposition in *Pennsylvania Coal Co. v. Mahon*, 260 U.S. 393, 67 L. Ed. 322, 43 S. Ct. 158 (1922), it was generally thought that the Takings Clause reached only a "direct appropriation" of property, Legal Tender Cases, 79 U.S. (12 Wall.) 457, 551, 20 L. Ed. 287 (1871), or the functional equivalent of a "practical ouster of [the owner's] possession," *Transportation Co. v. Chicago*, 99 U.S. 635, 642, 25 L. Ed. 336 (1879).... Justice Holmes recognized in Mahon, however, that if the protection against physical appropriations of private property was to be meaningfully enforced, the government's power to redefine the range of interests included in the ownership of property was necessarily constrained by constitutional limits. 260 U.S. at 414-415. If, instead, the uses of private property were subject to unbridled, uncompensated qualification under the police power, "the natural tendency of human nature [would be] to extend the qualification more and more until at last private property disappeared." Id., at 415. These considerations gave birth in that case to the oft-cited maxim that, "while property may be regulated to a certain extent, if regulation goes too far it will be recognized as a taking." Ibid.

Nevertheless, our decision in Mahon offered little insight into when, and under what circumstances, a given regulation would be seen as going "too far" for purposes of the Fifth Amendment. In 70-odd years of succeeding "regulatory takings" jurisprudence, we have generally eschewed any "'set formula'" for determining how far is too far, preferring to "engage in... essentially ad hoc, factual inquiries." *Penn Central Transportation Co. v. New York City*, 438 U.S. 104,...

We have, however, described at least two discrete categories of regulatory action as compensable without case-specific inquiry into the public interest advanced in support of the restraint. The first encompasses regulations that compel the property owner to suffer a physical "invasion" of his property. In general (at least with regard to permanent invasions), no matter how minute the intrusion, and no matter how weighty the public purpose behind it, we have required compensation. For example, in *Loretto v. Teleprompter Manhattan CATV Corp.*, 458 U.S. 419, 73 L. Ed. 2d 868, 102 S. Ct. 3164 (1982), we determined that New York's law requiring landlords to allow television cable companies to emplace cable facilities in their apartment buildings constituted a taking, id., at 435-440, even though the facilities occupied at most only 1 1/2 cubic feet of the landlords' property, see id., at 438, n.16....

The second situation in which we have found categorical treatment appropriate is where regulation denies all economically beneficial or productive use of land.... As we have said on numerous occasions, the Fifth Amendment is violated when land-use regulation "does not substantially advance legitimate state interests or denies an owner economically viable use of his land."...

We have never set forth the justification for this rule. Perhaps it is simply, as Justice Brennan suggested, that total deprivation of beneficial use is, from the landowner's point of view, the equivalent of a physical appropriation. See *San Diego Gas & Electric Co. v. San Diego*, 450 U.S. at 652 (dissenting opinion). "For what is the land but the profits thereof[?]" 1 E. Coke, *Institutes*, ch. 1, § 1 (1st Am. ed. 1812). Surely, at least, in the extraordinary circumstance when no productive or economically beneficial use of land is permitted, it is less realistic to indulge our usual assumption that the legislature is simply "adjusting the benefits and burdens of economic life," Penn Central Transportation Co., 438 U.S. at 124, in a manner that secures an "average reciprocity of advantage" to everyone concerned, *Pennsylvania Coal Co. v. Mahon*, 260 U.S. at 415. And the functional basis for permitting the government, by regulation, to affect property values without compensation -- that "Government hardly could go on if to some extent values incident to property could not be diminished without paying for every such change in the general law," id., at 413 -- does not apply to the relatively rare situations where the government has deprived a landowner of all economically beneficial uses.

On the other side of the balance, affirmatively supporting a compensation requirement, is the fact that regulations that leave the owner of land without economically beneficial or productive options for its use -- typically, as here, by requiring land to be left substantially in its natural state -- carry with them a heightened risk that private property is being pressed into some form of public service under the guise of mitigating serious public harm. See, e.g., *Annicelli v. South Kingstown,* 463 A.2d 133, 140-141 (R. I. 1983) (prohibition on construction adjacent to beach justified on twin grounds of safety and "conservation of open space"); *Morris County Land Improvement Co. v. Parsippany-Troy Hills Township,* 40 N.J. 539, 552-553, 193 A.2d 232, 240 (1963) (prohi-bition on filling marshlands imposed in order to preserve region as water detention basin and create wildlife refuge). As Justice Brennan explained: "From the government's point of view, the benefits flowing to the public from preservation of open space through regulation may be equally great as from creating a wildlife refuge through formal condemnation or increasing electricity production through a dam project that floods private property." San Diego Gas & Elec. Co., supra, at 652 (dissenting opinion). The many statutes o the books, both state and federal, that provide for the use of eminent domain to impose servitudes on private scenic lands preventing developmental uses, or to acquire such lands altogether, suggest the practical equivalence in this setting of negative regulation and appropriation....

We think, in short, that there are good reasons for our frequently expressed belief that when the owner of real property has been called upon to sacrifice all economically beneficial uses in the name of the common good, that is, to leave his property economically idle, he has suffered a taking.

B

The trial court found Lucas's two beachfront lots to have been rendered valueless by respondent's enforcement of the coastal-zone construction ban. Under Lucas's theory of the case, which rested upon our "no economically viable use" statements, that finding entitled him to compensation. Lucas believed it unnecessary to take issue with either the purposes behind the Beachfront Management Act, or the means chosen by the South Carolina Legislature to effectuate those purposes. The South Carolina Supreme Court, however, thought otherwise. In its view, the Beachfront Management Act was no ordinary enactment, but involved an exercise of South Carolina's "police powers" to mitigate the harm to the public interest that petitioner's use of his land might occasion. 304 S.C. at 384, 404 S.E.2d at 899. By neglecting to dispute the findings enumerated in the Act 10 or otherwise to challenge the legislature's purposes, petitioner "conceded that the beach/dune area of South Carolina's shores is an extremely valuable public resource; that the erection of new construction, inter alia, contributes to the erosion and destruction of this public resource; and that discouraging new construction in close proximity to the beach/dune area is necessary to prevent a great public harm." Id., at 382-383, 404 S.E.2d at 898. In the court's view, these concessions brought petitioner's challenge within a long line of this Court's cases sustaining against Due Process and Takings Clause challenges the State's use of its "police powers" to enjoin a property owner from activities akin to public nuisances. See *Mugler v. Kansas,* 123 U.S. 623, 31 L. Ed. 205, 8 S. Ct. 273 (1887) (law prohibiting manufacture of alcoholic beverages); *Hadacheck v. Sebastian,* 239 U.S. 394, 60 L. Ed. 348, 36 S. Ct. 143 (1915) (law barring operation of brick mill in residential area); *Miller v. Schoene,* 276 U.S. 272, 72 L. Ed. 568, 48 S. Ct. 246 (1928) (order to destroy diseased cedar trees to prevent infection of nearby orchards); *Goldblatt v. Hempstead,* 369 U.S. 590, 8 L. Ed. 2d 130, 82 S. Ct. 987 (1962) (law effectively preventing continued operation of quarry in residential area).

It is correct that many of our prior opinions have suggested that "harmful or noxious uses" of property may be proscribed by government regulation without the requirement of compensation. For a number of reasons, however, we think the South Carolina Supreme Court was too quick to conclude that that principle decides the present case. The "harmful or noxious uses" principle was the Court's early attempt to describe in theoretical terms why government may, consistent with the Takings Clause, affect property values by regulation without incurring an obligation to compensate -- a reality we nowadays acknowledge explicitly with respect to the full scope of the State's police power. See, e.g., Penn Central Transportation Co., 438 U.S. at 125 (where State "reasonably concludes that 'the health, safety, morals, or general welfare' would be promoted by prohibiting particular contemplated uses of land," compensation need not accompany prohibition); see also *Nollan v. California Coastal Comm'n,* 483 U.S. at 834-835 ("Our cases have not elaborated on the standards for determining what constitutes a 'legitimate state interest[,]' [but] they have made clear... that a broad range of governmental purposes and regulations satisfy these requirements"). We made this very point in Penn Central Transportation Co., where, in the course of sustaining New York City's landmarks preservation program against a takings challenge, we rejected the petitioner's suggestion that Mugler and the cases following it were premised on, and thus limited by, some objective conception of "noxiousness":

> "The uses in issue in Hadacheck, Miller, and Goldblatt were perfectly lawful in themselves. They involved no 'blameworthiness,... moral wrongdoing or conscious act of dangerous risk-taking which induced society to shift the cost to a particular individual.' Sax, 'Takings and the Police Power,' 74 *Yale L. J.* 36, 50 (1964). These cases are better understood as resting not on any supposed 'noxious' quality of the prohibited uses but rather on the ground that the restrictions were reasonably related to the implementation of a policy -- not unlike historic preservation -- expected to produce a widespread public benefit and applicable to all similarly situated property." 438 U.S. at 133-134, n.30.

"Harmful or noxious use" analysis was, in other words, simply the progenitor of our more contemporary statements that "land-use regulation does not effect a taking if it 'substantially advances legitimate state interests'...." Nollan, supra, at 834 (quoting *Agins v. Tiburon*, 447 U.S. at 260....

The transition from our early focus on control of "noxious" uses to our contemporary understanding of the broad realm within which government may regulate without compensation was an easy one, since the distinction between "harm-preventing" and "benefit-conferring" regulation is often in the eye of the beholder. It is quite possible, for example, to describe in either fashion the ecological, economic, and esthetic concerns that inspired the South Carolina Legislature in the present case. One could say that imposing a servitude on Lucas's land is necessary in order to prevent his use of it from "harming" South Carolina's ecological resources; or, instead, in order to achieve the "benefits" of an ecological preserve. 11 Compare, e.g., *Claridge v. New Hampshire Wetlands Board,* 125 N.H. 745, 752, 485 A.2d 287, 292 (1984) (owner may, without compensation, be barred from filling wetlands because landfilling would deprive adjacent coastal habitats and marine fisheries of ecological support), with, e.g., *Bartlett v. Zoning Comm'n of Old Lyme,* 161 Conn. 24, 30, 282 A.2d 907, 910 (1971) (owner barred from filling tidal marshland must be compensated, despite municipality's "laudable" goal of "preserving marshlands from encroachment or destruction"). Whether one or the other of the competing characterizations will come to one's lips in a particular case depends primarily upon one's evaluation of the worth of competing uses of real estate.... A given restraint will be seen as mitigating "harm" to the adjacent parcels or securing a "benefit" for them, depending upon the observer's evaluation of the relative importance of the use that the restraint favors.... Whether Lucas's construction of single-family residences on his parcels should be described as bringing "harm" to South Carolina's adjacent ecological resources thus depends principally upon whether the describer believes that the State's use interest in nurturing those resources is so important that any competing adjacent use must yield.

When it is understood that "prevention of harmful use" was merely our early formulation of the police power justification necessary to sustain (without compensation) any regulatory diminution in value; and that the distinction between regulation that "prevents harmful use" and that which "confers benefits" is difficult, if not impossible, to discern on an objective, value-free basis; it becomes self-evident that noxious-use logic cannot serve as a touchstone to distinguish regulatory "takings" -- which require compensation -- from regulatory deprivations that do not require compensation. A fortiori the legislature's recitation of a noxious-use justification cannot be the basis for departing from our categorical rule that total regulatory takings must be compensated. If it were, departure would virtually always be allowed. The South Carolina Supreme Court's approach would essentially nullify Mahon's affirmation of limits to the noncompensable exercise of the police power. Our cases provide no support for this: None of them that employed the logic of "harmful use" prevention to sustain a regulation involved an allegation that the regulation wholly eliminated the value of the claimant's land....

Where the State seeks to sustain regulation that deprives land of all economically beneficial use, we think it may resist compensation only if the logically antecedent inquiry into the nature of the owner's estate shows that the proscribed use interests were not part of his title to begin with. This accords, we think, with our "takings" jurisprudence, which has traditionally been guided by the understandings of our citizens regarding the content of, and the State's power over, the "bundle of rights" that they acquire when they obtain title to property. It seems to us that the property owner necessarily expects the uses of his property to be restricted, from time to time, by various measures newly enacted by the State in legitimate exercise of its police powers; "as long recognized, some values are enjoyed under an implied limitation and must yield to the police power." *Pennsylvania Coal Co. v. Mahon,* 260 U.S. at 413. And in the case of personal property, by reason of the State's traditionally high degree of control over commercial dealings, he ought to be aware of the possibility that new regulation might even render his property economically worthless (at least if the property's only economically productive use is sale or manufacture for sale). See *Andrus v. Allard,* 444 U.S. 51, 66-67, 62 L. Ed. 2d 210, 100 S. Ct. 318 (1979) (prohibition on sale of eagle feathers). In the case of land, however, we think the notion pressed by the Council that title is somehow held subject to the "implied limitation" that the State may subsequently eliminate all economically valuable use is inconsistent with the historical compact recorded in the Takings Clause that has become part of our constitutional culture.

Where "permanent physical occupation" of land is concerned, we have refused to allow the government to decree it anew (without compensation), no matter how weighty the asserted "public interests" involved, *Loretto v. Teleprompter Manhattan CATV Corp.,* 458 U.S. at 426 — though we assuredly would permit the government to assert a permanent easement that was a pre-existing limitation upon the landowner's title.... We believe similar treatment must be accorded confiscatory regulations, i.e., regulations that prohibit all economically beneficial use of land: Any limitation so severe cannot be newly legislated or decreed (without compensation), but must inhere in the title itself, in the restrictions that background principles of the State's law of property and nuisance already place upon land ownership. A law or decree with such an effect must, in other words, do no more than duplicate the result that could have been achieved in the courts -- by adjacent landowners (or other uniquely affected persons) under the State's law of private nuisance, or by the State under its complementary power to abate nuisances that affect the public generally, or otherwise.

On this analysis, the owner of a lakebed, for example, would not be entitled to compensation when he is denied the requisite permit to engage in a landfilling operation that would have the effect of flooding others' land. Nor the corporate owner of a nuclear generating plant, when it is directed to remove all improvements from its land upon discovery that the plant sits astride an earthquake fault. Such regulatory action may well have the effect of eliminating the land's only economically productive use, but it does not proscribe a productive use that was previously permissible under relevant property and nuisance principles. The use of these properties for what are now expressly prohibited purposes was always unlawful, and (subject to other constitutional limitations) it was open to the State at any point to make the implication of those background principles of nuisance and property law explicit.... In light of our traditional resort to "existing rules or understandings that stem from an independent source such as state law" to define the range of interests that qualify for protection as "property" under the Fifth and Fourteenth Amendments,...

The "total taking" inquiry we require today will ordinarily entail (as the application of state nuisance law ordinarily entails) analysis of, among other things, the degree of harm to public lands and resources, or adjacent private property, posed by the claimant's proposed activities... the social value of the claimant's activities and their suitability to the locality in question..., and the relative ease with which the alleged harm can be avoided through measures taken by the claimant and the government (or adjacent private landowners) alike.... The fact that a particular use has long been engaged in by similarly situated owners ordinarily imports a lack of any common-law prohibition (though changed circumstances or new knowledge may make what was previously permissible no longer so, see id., § 827, Comment g. So also does the fact that other landowners, similarly situated, are permitted to continue the use denied to the claimant.

It seems unlikely that common-law principles would have prevented the erection of any habitable or productive improvements on petitioner's land; they rarely support prohibition of the "essential use" of land, *Curtin v. Benson*, 222 U.S. 78, 86, 56 L. Ed. 102, 32 S. Ct. 31 (1911).The question, however, is one of state law to be dealt with on remand. We emphasize that to win its case South Carolina must do more than proffer the legislature's declaration that the uses Lucas desires are inconsistent with the public interest, or the conclusory assertion that they violate a common-law maxim such as sic utere tuo ut alienum non laedas. As we have said, a "State, by ipse dixit, may not transform private property into public property without compensation... ." *Webb's Fabulous Pharmacies, Inc. v. Beckwith*, 449 U.S. 155, 164, 66 L. Ed. 2d 358, 101 S. Ct. 446 (1980). Instead, as it would be required to do if it sought to restrain Lucas in a common-law action for public nuisance, South Carolina must identify background principles of nuisance and property law that prohibit the uses he now intends in the circumstances in which the property is presently found. Only on this showing can the State fairly claim that, in proscribing all such beneficial uses, the Beachfront Management Act is taking nothing.

* * *

The judgment is reversed, and the case is remanded for proceedings not inconsistent with this opinion.

So ordered.

[T]o provide for mutual assistance between the states entering into this compact

---

JOINT RESOLUTION CONSENTING TO EMERGENCY MANAGEMENT ASSISTANCE COMPACT

1996

Introduction

The Constitution permits states to enter into interstate compacts with congressional approval. On October 19, 1996, Congress adopted this resolution affirming a compact that states, mostly in the South, had entered into for mutual aid. Contiguous states are often in the best position to help one another, and this resolution affirms this.

Public Law 104-321, 104th Congress
Joint Resolution

Granting the consent of Congress to the Emergency Management Assistance Compact.

Resolved by the Senate and House of Representatives of the United States of America in Congress assembled,

SECTION 1. CONGRESSIONAL CONSENT

The Congress consents to the Emergency Management Assistance Compact entered into by Delaware, Florida, Georgia, Louisiana, Maryland, Mississippi, Missouri, Oklahoma, South Carolina, South Dakota, Tennessee, Virginia, and West Virginia. The compact reads substantially as follows:

Emergency Management Assistance Compact

ARTICLE I
PURPOSE AND AUTHORITIES

This compact is made and entered into by and between the participating member states which enact this compact...

The purpose of this compact is to provide for mutual assistance between the states entering into this compact in managing any emergency disaster that is duly declared by the Governor of the affected states, whether arising from natural disaster, technological hazard, man-made disaster, civil emergency aspects of resources shortages, community disorders, insurgency, or enemy attack.

This compact shall also provide for mutual cooperation in emergency-related exercises, testing, other training activities using equipment and personnel simulating performance of any aspects of the giving and receiving of aid by party states or subdivisions of party states during emergencies, such actions occurring outside actual declared emergency periods. Mutual assistance in this compact may include the use of the states' National Guard forces, either in accordance with the National Guard Mutual Assistance Compact or by mutual agreement between states.

ARTICLE II.
GENERAL IMPLEMENTATION

Each party state entering into this compact recognizes that many emergencies transcend political jurisdictional boundaries and that intergovernmental coordination is essential in managing these and other emergencies under this compact. Each state further recognizes that there will be emergencies which require immediate access and present procedures to apply outside resources to make a prompt and effective response to such an emergency. This is because few, if any, individual states have all the resources they may need in all types of emergencies or the capability of delivering resources to areas where emergencies exist.

The prompt, full, and effective utilization of resources of the participating states, including any resources on hand or available from the federal government or any other source, that are essential to the safety, care, and welfare of the people in the event of any emergency or disaster declared by a party state, shall be the underlying principle on which all articles of this compact shall be understood.

On behalf of the Governor of each state participating in the compact, the legally designated state official who is assigned responsibility for emergency management will be responsible for formulation of the appropriate interstate mutual aid plans and procedures necessary to implement this compact.

ARTICLE IV.
LIMITATIONS

Any party state requested to render mutual aid or conduct exercises and training for mutual aid shall take such action as is necessary to provide and make available the resources covered by this compact in accordance with the terms hereof; provided that it is understood that the state rendering aid may without resources to the extent necessary to provide reasonable protection for such state.

Each party state shall afford to the emergency forces of any party state, while operating within its state limits under the terms and conditions of this compact, the same power except that of arrest unless specifically authorized by the receiving state, duties, rights, and privileges as are afforded forces of the state in which they are performing emergency services. Emergency forces will continue under the command and control of their regular leaders, but the organizational units will come under the operational control of the emergency services authorities of the state receiving assistance...

ARTICLE X.
EVACUATION

Plans for the orderly evacuation and interstate reception of portions of the civilian population as the result of any emergency or disaster of sufficient proportions to so warrant, shall be worked out and maintained between the party states and the emergency management/services directors of the various jurisdictions where any type of incident requiring evacuations might occur. Such plans shall be put into effect by request of the state from which evacuees come and shall include the manner of transporting such evacuees, the number of evacuees to be received in different areas, the manner in which food, clothing, housing, and medical care will be provided, the registration of the evacuees, the providing of facilities for the notification of relatives or friends, and the forwarding of such to other areas or the bringing in of additional materials, supplies, and all other relevant factors. Such plans shall provide that the party states receiving evacuees and the party state from which the evacuees come shall mutually agree as to reimbursement of out-of-pocket expenses incurred in receiving and caring for such evacuees, for expenditures for transportation, food, clothing, medicines, and medical care, and like items. Such expenditures shall be reimbursed as agreed by the party state from which the evacuees come. After the termination of the emergency or disaster, the party state from which the evacuees come shall assume the responsibility for the ultimate support of repatriation of such evacuees.

[A] program for predisaster mitigation

DISASTER MITIGATION ACT OF 2000

(December 6)

Introduction

As the number of disasters, and disaster declarations, began to rise, there was increased concern that the government should be devoting greater attention to disaster prevention. This law, only a small part of which is excerpted here, was designed to encourage states to put mitigation procedures in place as a condition of receiving federal aid.

AN ACT

To amend the Robert T. Stafford Disaster Relief and Emergency Assistance Act to authorize a program for predisaster mitigation, to streamline the administration of disaster relief, to control the federal costs of disaster assistance, and for other purposes.

Be it enacted by the Senate and House of Representatives of the United States of America in Congress assembled

SEC. 101. FINDINGS AND PURPOSE

(a) FINDINGS - Congress finds that—

(1) natural disasters, including earthquakes, tsunamis, tornadoes, hurricanes, flooding, and wildfires, pose great danger to human life and to property throughout the United States;

(2) greater emphasis needs to be placed on—

(A) identifying and assessing the risks to States and local governments (including Indian tribes) from natural disasters;

(B) implementing adequate measures to reduce losses from natural disasters; and

(C) ensuring that the critical services and facilities of communities will continue to function after a natural disaster;

(3) expenditures for postdisaster assistance are increasing without commensurate reductions in the likelihood of future losses from natural disasters;

(4) in the expenditure of Federal funds under the Robert T. Stafford Disaster Relief and Emergency Assistance Act (42 U.S.C. 5121 et seq.), high priority should be given to mitigation of hazards at the local level; and

(5) with a unified effort of economic incentives, awareness and education, technical assistance, and demonstrated Federal support, States and local governments (including Indians tribes) will be able to—

(A) form effective community-based partnerships for hazard mitigation purposes;

(B) implement effective hazard mitigation measures that reduce the potential damage from natural disasters;

(C) ensure continued functionality of critical ser ices;

(D) leverage additional non-Federal resources in meeting natural disaster resistance goals; and

(E) make commitments to long-term hazard mitigation efforts to be applied to new and existing structures.

(b) PURPOSE- The purpose of this title is to establish a national disaster hazard mitigation program--

(1) To reduce the loss of life and property, human suffering, economic disruption, and disaster assistance costs resulting from natural disasters; and

(2) To provide a source of predisaster hazard mitigation funding that will assist States and local governments (including Indian tribes) in implementing effective hazard mitigation measures that are designed to ensure the continued functionality of critical services and facilities after a natural disaster.

SEC. 322. MITIGATION PLANNING

(a) REQUIRMENT OF MITIGATION PLAN- As a condition of receipt of an increased Federal share for hazard mitigation measures under subsection €, a State, local, or tribal government shall develop and submit for approval to the President, a mitigation plan that outlines processes for identifying the natural hazards, risks, and vulnerabilities of the area under the jurisdiction of the government.

(c) STATE PLANS – The State process of development of a mitigation plan under this section shall—

(1) Identify the natural hazards, risks, and vulnerabilities of areas in the State;

(2) Support development of local mitigation plans;

(3) Provide for technical assistance to local and tribal governments for mitigation planning; and

(4) Identify and prioritize mitigation actions that the State will support, as resources become available.

SEC. 408. FEDERAL ASSISTANCE TO INDIVIDUALS AND HOUSEHOLDS

(a) IN GENERAL-

(1) PROVISIONS OF ASSISTANCE- In accordance with this section, the President, in consultation with the Governor of a State, may provide financial assistance, and, if necessary, direct services, to individuals and households in the State, who, as a direct result of a major disaster, have necessary expenses and serious needs in cases in which the individuals and households are unable to meet such expenses or needs through other means.

(b) HOUSING ASSISTANCE

(1) Eligibility-The President may provide financial or other assistance under this section to individuals and households to respond to the disaster-related housing needs of individuals and households who are displaced from their predisaster primary residences or whose predisaster primary residences are rendered uninhabitable as a result of damage caused by a major disaster.

Source: https://www.govtrack.us/congress/bills/016/hr707/text. Accessed 12/4/2016.

PART VII

9/11, Hurricane Katrina, Global Warming, and Recent Disasters

Whether we bring our enemies to justice or bring justice to our enemies, justice will be done

---

TRANSCRIPT OF PRESIDENT BUSH'S ADDRESS TO A JOINT SESSION OF CONGRESS

September 20, 2001

Introduction

Just as the Japanese attack on Pearl Harbor on December 7, 1941 was a defining event for that generation of Americans, so too, the attacks on the World Trade Center in New York City and the Pentagon outside Washington, D.C. proved to be an axial moment for Americans of that time. The "war on terrorism" seemed much more urgent than most efforts designed to mitigate natural disasters. The U.S. subsequently engaged in wars in Afghanistan and Iraq to root out terrorism, with other actions in Syria and neighboring countries. The attacks resulted in the creation of the Department of Homeland Security, which was elevated to cabinet status and resulted in significant governmental expenditures to rebuild New York City.

Mr. Speaker, Mr. President Pro Tempore, members of Congress, and fellow Americans, in the normal course of events, presidents come to this chamber to report on the state of the union. Tonight, no such report is needed; it has already been delivered by the American people.

We have seen it in the courage of passengers who rushed terrorists to save others on the ground. Passengers like an exceptional man named Todd Beamer. And would you please help me welcome his wife Lisa Beamer here tonight?

We have seen the state of our union in the endurance of rescuers working past exhaustion.

We've seen the unfurling of flags, the lighting of candles, the giving of blood, the saying of prayers in English, Hebrew and Arabic.

We have seen the decency of a loving and giving people who have made the grief of strangers their own.

My fellow citizens, for the last nine days, the entire world has seen for itself the state of union, and it is strong.

Tonight, we are a country awakened to danger and called to defend freedom. Our grief has turned to anger and anger to resolution. Whether we bring our enemies to justice or bring justice to our enemies, justice will be done.

I thank the Congress for its leadership at such an important time.

All of America was touched on the evening of the tragedy to see Republicans and Democrats joined together on the steps of this Capitol singing "God Bless America."

And you did more than sing. You acted, by delivering $40 billion to rebuild our communities and meet the needs of our military. Speaker Hastert, Minority Leader Gephardt, Majority Leader Daschle and Senator Lott, I thank you for your friendship, for your leadership and for your service to our country.

And on behalf of the American people, I thank the world for its outpouring of support.

America will never forget the sounds of our national anthem playing at Buckingham Palace, on the streets of Paris and at Berlin's Brandenburg Gate.

We will not forget South Korean children gathering to pray outside our embassy in Seoul, or the prayers of sympathy offered at a mosque in Cairo.

We will not forget moments of silence and days of mourning in Australia and Africa and Latin America.

Nor will we forget the citizens of 80 other nations who died with our own. Dozens of Pakistanis, more than 130 Israelis, more than 250 citizens of India, men and women from El Salvador, Iran, Mexico and Japan, and hundreds of British citizens.

America has no truer friend than Great Britain. Once again, we are joined together in a great cause.

I'm so honored the British prime minister has crossed an ocean to show his unity with America.

Thank you for coming, friend.

On September the 11th, enemies of freedom committed an act of war against our country. Americans have known wars, but for the past 136 years they have been wars on foreign soil, except for one Sunday in 1941. Americans have known the casualties of war, but not at the center of a great city on a peaceful morning.

Americans have known surprise attacks, but never before on thousands of civilians. All of this was brought upon us in a single day, and night fell on a different world, a world where freedom itself is under attack.

Americans have many questions tonight. Americans are asking, "Who attacked our country?"

The evidence we have gathered all points to a collection of loosely affiliated terrorist organizations known as al Qaeda. They are some of the murderers indicted for bombing American embassies in Tanzania and Kenya and responsible for bombing the USS Cole.

Al Qaeda is to terror what the Mafia is to crime. But its goal is not making money, its goal is remaking the world and imposing its radical beliefs on people everywhere.

The terrorists practice a fringe form of Islamic extremism that has been rejected by Muslim scholars and the vast majority of Muslim clerics; a fringe movement that perverts the peaceful teachings of Islam.

The terrorists' directive commands them to kill Christians and Jews, to kill all Americans and make no distinctions among military and civilians, including women and children. This group and its leader, a person named Osama bin Laden, are linked to many other organizations in different countries, including the Egyptian Islamic Jihad, the Islamic Movement of Uzbekistan.

There are thousands of these terrorists in more than 60 countries.

They are recruited from their own nations and neighborhoods and brought to camps in places like Afghanistan where they are trained in the tactics of terror. They are sent back to their homes or sent to hide in countries around the world to plot evil and destruction. The leadership of al Qaeda has great influence in Afghanistan and supports the Taliban regime in controlling most of that country. In Afghanistan we see al Qaeda's vision for the world. Afghanistan's people have been brutalized, many are starving and many have fled.

Women are not allowed to attend school. You can be jailed for owning a television. Religion can be practiced only as their leaders dictate. A man can be jailed in Afghanistan if his beard is not long enough. The United States respects the people of Afghanistan—after all, we are currently its largest source of humanitarian aid—but we condemn the Taliban regime.

It is not only repressing its own people, it is threatening people everywhere by sponsoring and sheltering and supplying terrorists.

By aiding and abetting murder, the Taliban regime is committing murder. And tonight the United States of America makes the following demands on the Taliban:

–Deliver to United States authorities all of the leaders of Al Qaeda who hide in your land.

–Release all foreign nationals, including American citizens you have unjustly imprisoned.

–Protect foreign journalists, diplomats and aid workers in your country.

–Close immediately and permanently every terrorist training camp in Afghanistan. And hand over every terrorist and every person and their support structure to appropriate authorities.

–Give the United States full access to terrorist training camps, so we can make sure they are no longer operating.

These demands are not open to negotiation or discussion.

The Taliban must act and act immediately.

They will hand over the terrorists or they will share in their fate. I also want to speak tonight directly to Muslims throughout the world. We respect your faith. It's practiced freely by many millions of Americans and by millions more in countries that America counts as friends. Its teachings are good and peaceful, and those who commit evil in the name of Allah blaspheme the name of Allah.

The terrorists are traitors to their own faith, trying, in effect, to hijack Islam itself.

The enemy of America is not our many Muslim friends. It is not our many Arab friends. Our enemy is a radical network of terrorists and every government that supports them.

Our war on terror begins with al Qaeda, but it does not end there.

It will not end until every terrorist group of global reach has been found, stopped and defeated.

Americans are asking "Why do they hate us?"

They hate what they see right here in this chamber: a democratically elected government. Their leaders are self-appointed. They hate our freedoms: our freedom of religion, our freedom of speech, our freedom to vote and assemble and disagree with each other.

They want to overthrow existing governments in many Muslim countries such as Egypt, Saudi Arabia and Jordan. They want to drive Israel out of the Middle East. They want to drive Christians and Jews out of vast regions of Asia and Africa.

These terrorists kill not merely to end lives, but to disrupt and end a way of life. With every atrocity, they hope that America grows fearful, retreating from the world and forsaking our friends. They stand against us because we stand in their way.

We're not deceived by their pretenses to piety.

We have seen their kind before. They're the heirs of all the murderous ideologies of the 20th century. By sacrificing human life to serve their radical visions, by abandoning every value except the will to power, they follow in the path of fascism, Nazism and totalitarianism. And they will follow that path all the way to where it ends in history's unmarked grave of discarded lies. Americans are asking, "How will we fight and win this war?"

We will direct every resource at our command—every means of diplomacy, every tool of intelligence, every instrument of law enforcement, every financial influence, and every necessary weapon of war—to the destruction and to the defeat of the global terror network.

Now, this war will not be like the war against Iraq a decade ago, with a decisive liberation of territory and a swift conclusion. It will not look like the air war above Kosovo two years ago, where no ground troops were used and not a single American was lost in combat.

Our response involves far more than instant retaliation and isolated strikes. Americans should not expect one battle, but a lengthy campaign unlike any other we have ever seen. It may include dramatic strikes visible on TV and covert operations secret even in success.

We will starve terrorists of funding, turn them one against another, drive them from place to place until there is no refuge or no rest.

And we will pursue nations that provide aid or safe haven to terrorism. Every nation in every region now has a decision to make: Either you are with us or you are with the terrorists.

From this day forward, any nation that continues to harbor or support terrorism will be regarded by the United States as a hostile regime. Our nation has been put on notice, we're not immune from attack. We will take defensive measures against terrorism to protect Americans. Today, dozens of federal departments and agencies, as well as state and local governments, have responsibilities affecting homeland security.

These efforts must be coordinated at the highest level. So tonight, I announce the creation of a Cabinet-level position reporting directly to me, the Office of Homeland Security. And tonight, I also announce a distinguished American to lead this effort, to strengthen American security: a military veteran, an effective governor, a true patriot, a trusted friend, Pennsylvania's Tom Ridge.

He will lead, oversee and coordinate a comprehensive national strategy to safeguard our country against terrorism and respond to any attacks that may come. These measures are essential. The only way to defeat terrorism as a threat to our way of life is to stop it, eliminate it and destroy it where it grows.

Many will be involved in this effort, from FBI agents, to intelligence operatives, to the reservists we have called to active duty. All deserve our thanks, and all have our prayers. And tonight a few miles from the damaged Pentagon, I have a message for our military: Be ready. I have called the armed forces to alert, and there is a reason.

The hour is coming when America will act, and you will make us proud.

This is not, however, just America's fight. And what is at stake is not just America's freedom. This is the world's fight. This is civilization's fight. This is the fight of all who believe in progress and pluralism, tolerance and freedom.

We ask every nation to join us.

We will ask and we will need the help of police forces, intelligence service and banking systems around the world. The United States is grateful that many nations and many international organizations have already responded with sympathy and with support—nations from Latin America to Asia to Africa to Europe to the Islamic world.

Perhaps the NATO charter reflects best the attitude of the world: An attack on one is an attack on all. The civilized world is rallying to America's side.

They understand that if this terror goes unpunished, their own cities, their own citizens may be next. Terror unanswered can not only bring down buildings, it can threaten the stability of legitimate governments.

And you know what? We're not going to allow it.

Americans are asking, "What is expected of us?"

I ask you to live your lives and hug your children. I know many citizens have fears tonight, and I ask you to be calm and resolute, even in the face of a continuing threat.

I ask you to uphold the values of America and remember why so many have come here.

We're in a fight for our principles, and our first responsibility is to live by them. No one should be singled out for unfair treatment or unkind words because of their ethnic background or religious faith.

I ask you to continue to support the victims of this tragedy with your contributions. Those who want to give can go to a central source of information, Libertyunites.org, to find the names of groups providing direct help in New York, Pennsylvania and Virginia. The thousands of FBI agents who are now at work in this investigation may need your cooperation, and I ask you to give it. I ask for your patience with the delays and inconveniences that may accompany tighter security and for your patience in what will be a long struggle.

I ask your continued participation and confidence in the American economy. Terrorists attacked a symbol of American prosperity; they did not touch its source.

America is successful because of the hard work and creativity and enterprise of our people. These were the true strengths of our economy before September 11, and they are our strengths today.

And finally, please continue praying for the victims of terror and their families, for those in uniform and for our great country. Prayer has comforted us in sorrow and will help strengthen us for the journey ahead. Tonight I thank my fellow Americans for what you have already done and for what you will do.

And ladies and gentlemen of the Congress, I thank you, their representatives, for what you have already done and for what we will do together.

Tonight we face new and sudden national challenges. We will come together to improve air safety, to dramatically expand the number of air marshals on domestic flights and take new measures to prevent hijacking.

We will come together to promote stability and keep our airlines flying with direct assistance during this emergency.

We will come together to give law enforcement the additional tools it needs to track down terror here at home.

We will come together to strengthen our intelligence capabilities to know the plans of terrorists before they act and to find them before they strike.

We will come together to take active steps that strengthen America's economy and put our people back to work.

Tonight, we welcome two leaders who embody the extraordinary spirit of all New Yorkers, Governor George Pataki and Mayor Rudolph Giuliani.

As a symbol of America's resolve, my administration will work with Congress and these two leaders to show the world that we will rebuild New York City.

After all that has just passed, all the lives taken and all the possibilities and hopes that died with them, it is natural to wonder if America's future is one of fear.

Some speak of an age of terror. I know there are struggles ahead and dangers to face. But this country will define our times, not be defined by them.

As long as the United States of America is determined and strong, this will not be an age of terror. This will be an age of liberty here and across the world.

Great harm has been done to us. We have suffered great loss. And in our grief and anger we have found our mission and our moment.

Freedom and fear are at war. The advance of human freedom, the great achievement of our time and the great hope of every time, now depends on us.

Our nation, this generation, will lift the dark threat of violence from our people and our future. We will rally the world to this cause by our efforts, by our courage. We will not tire, we will not falter and we will not fail.

It is my hope that in the months and years ahead life will return almost to normal. We'll go back to our lives and routines and that is good.

Even grief recedes with time and grace.

But our resolve must not pass. Each of us will remember what happened that day and to whom it happened. We will remember the moment the news came, where we were and what we were doing.

Some will remember an image of a fire or story or rescue. Some will carry memories of a face and a voice gone forever.

And I will carry this. It is the police shield of a man named George Howard who died at the World Trade Center trying to save others.

It was given to me by his mom, Arlene, as a proud memorial to her son. It is my reminder of lives that ended and a task that does not end.

I will not forget the wound to our country and those who inflicted it. I will not yield, I will not rest, I will not relent in waging this struggle for freedom and security for the American people. The course of this conflict is not known, yet its outcome is certain. Freedom and fear, justice and cruelty, have always been at war, and we know that God is not neutral between them.

Fellow citizens, we'll meet violence with patient justice, assured of the rightness of our cause and confident of the victories to come.

In all that lies before us, may God grant us wisdom and may he watch over the United States of America. Thank you.

Source: Transcript of President Bush's address, CNN.com. http://xcnn.allpolitics.printthis.clickability.com/pt/cpt?expire=-1&title=CNN.com+-+Trans... 11/26/2016.

Note: Applause notations have been removed.

Acting as a focal point regarding natural and manmade crises and emergency planning

---

PUBLIC LAW ESTABLISHING DEPARTMENT OF HOMELAND SECURITY

107-296, November 25, 2002, 116 Stat. 2135, 107th Congress

Introduction

No event in modern U.S. history has arguably been any more traumatic than the terrorist attacks by Islamic extremists on the homeland on September 11, 2001 that resulted in the destruction of the twin towers in New York City and massive loss of life. Congress responded almost immediately with legislation, including this law, which created the Department of Homeland Security and elevated its director to a cabinet post. As proved to be the case during the Cold War, preparations for and reactions to natural disasters often have to be balanced against external threats. Notably, foreign terrorism was often tied financially to drug cartels, which also posed national security threats.

AN ACT

To establish the Department of Homeland Security, and for other purposes.

Sec. 101. EXECUTIVE DPEARTMENT; MISSION.

(1) IN GENERAL.—There is established a Department of Homeland Security, as an executive department of the United States within the meaning of title 5, United States Code.

(b) MISSION.—

(1) IN GENERAL. –The primary mission of the Department is to—

(A) prevent terrorist attacks within the United States;

(B) reduce the vulnerability of the United States to terrorism;

(C) minimize the damage, and assist in the recovery, from terrorist attacks that do occur within the United States

(D) carry out all functions of entities transferred to the Department, including by acting as a focal point regarding natural and manmade crises and emergency planning;

(E) ensure that the functions of the agencies and subdivisions within the Department that are not related directly to securing the homeland are not diminished or neglected expect by a specific explicit Act of congress;

(F) ensure that the overall economic security of the United States is not diminished by efforts, activities, and programs aimed at securing the homeland; and

(G) monitor connections between illegal drug trafficking and terrorism, coordinate efforts to sever such connections, and otherwise contribute to efforts to interdict illegal drug trafficking.

(2) RESPONSIBILOITY FOR INVESTIGATING AND PROSECUTING TERRORISM.— Except as specifically provided by law with respect to entities transferred to the Department under this Act, primary responsibility for investigating and prosecuting acts of terrorism shall be vested not in the Department, but rather in Federal, State, and local law enforcement agencies with jurisdiction over the acts in question.

Sec. 102. SECRETARY; FUNCTIONS.

(a) SECRETARY. -

(1) IN GENERAL. -There is a Secretary of Homeland Security, appointed by the President, by and with the advice and consent of the Senate.

(2) HEAD OF DEPARTMENT. - The Secretary is the head of the Department and shall have direction, authority, and control over it.

[T]o establish a comprehensive, national, all-hazards approach to domestic incident management

---

NATIONAL RESPONSE PLAN

2004

Introduction

This 2004 Plan was intended to provide a template for governmental responses to extraordinary events, which the President could designate as Incidents of National Significance. These protocols were designed both for natural disasters and for terrorism and other national security threats. Although they have been updated, such protocols remain in place.

Purpose

The purpose of the NRP is to establish a comprehensive, national, all-hazards approach to domestic incident management across a spectrum of activities including prevention, preparedness, response, and recovery. (p. 2)

Scope and Applicability

This plan distinguishes between incidents that require DHS coordination, termed incidents of National Significance, and the majority of incidents occurring each year that are handled by responsible jurisdictions or agencies through other established authorities and existing plans. (p. 3)

Incidents of National Significance

The NRP bases the definition of Incidents of National Significance on situations related to the following four criteria, set forth in HSPD-5:

1. A Federal department or agency acting under its own authority has requested the assistance of the Secretary of Homeland Security.

2. The resources of State and local authorities are overwhelmed and Federal assistance has been requested by the appro0priate State and local authorities. Examples include:

Major disasters or emergencies as defined under the Stafford Act; and Catastrophic incidents ...

3. More than one Federal department or agency has become substantially involved in responding to an incident.

Examples include:

Credible threats, indications or warnings or imminent terrorist attack, or acts of terrorism directed domestically against the people, property, environment, or political or legal institutions of the United States or its territories or possessions; and

Threats or incidents related to high-profile, large-scale events that present high-probability targets, such as National Special Security Events NSSEs) and other special events as determined by the Secretary of Homeland Security, in coordination with other Federal departments and agencies.

4. The Secretary of Homeland Security has been directed to assume responsibility for managing a domestic incident by the President. (p. 4)

II. Planning Assumptions & Considerations

The NRP is based on the planning assumptions and considerations presented in this section.

Incidents are typically managed at the lowest possible geographic, organization, and jurisdictional level.

Incident management activities will be initiated and conducted using the principles contained in the NIMS.

The combined expertise and capabilities of government at all levels, the private sector, and nongovernmental organizations most likely will be required to prevent, prepare for, respond to, and recover for Incidents of National Significance.

Incidents of National Significance require the Secretary of Homeland Security to coordinate operations and/or resources, and may:

Occur at any time with little or no warning in the context of a general or specific threat or hazard;

Require significant information-sharing at the unclassified and classified levels across multiple jurisdictions and between the public and private sectors;

Involve single or multiple geographic areas;

Have significant international impact and/or require significant international information-sharing, resource coordination, and/or assistance;

Span the spectrum of incident management to include prevention, preparedness, response, and recovery;

Involve multiple, highly varied hazards or threats on a local, regional, or national scale;

Result in numerous casualties; fatalities; displaced people; property loss; distruption of normal life-support systems, essential public services and basic infrastructure; and significant damage to the environment;

Impact critical infrastructures across sectors;

Overwhelm capabilities of State, local, and tribal governments and private-sector infrastructure owners and operators;

Attract a sizeable influx of independent, spontaneous volunteers and supplies;

Require extremely short-notice Federal asset coordination and response timelines; and

Require prolonged, sustained incident management operations and support activities.

Top priorities of incident management are to:

Save lives and protect the health and safety of the public, responders, and recovery workers;

Ensure security of the homeland;

Prevent an imminent incident, including acts of terrorism, from occurring;

Protect and restore critical infrastructure and key resources;

Conduct law enforcement investigations to resolve the incident, apprehend the perpetrators, and collect and preserve evidence for prosecution and/or attribution;

Protect property and mitigate damages and impacts to individuals, communities, and the environment; and

Facilitate recovery of individuals, families, businesses, governments and the environment.... (p. 6)

Under provisions of the Stafford Act and applicable regulations:

> A Governor may request the President to declare a major disaster or emergency if the Governor finds that effective response to the event is beyond the combined response capabilities of the State and affected local governments. Based on the findings of a joint Federal-State-local Preliminary Damage Assessment (PDA) indicating the damages are of sufficient severity and magnitude to warrant assistance under the act, the President may grant a major disaster or emergency declaration. (Note: In a particularly fast-moving or clearly devastating disaster, the PDA process may be deferred until after the declaration.)
>
> If the President determines that an emergency exists where the primary responsibility for response rests with the Government of the United States, or because the emergency involves an area or facility for which the Federal Government exercises exclusive or preeminent primary responsibility and authority, the President may unilaterally direct the provision of assistance under the act and will, if practicable, consult with the Governor of the State.
>
> DHS can use limited pre-declaration authorities to move initial response resources (critial goods typically needed in the immediate aftermath of a disaster such as food, water, emergency generators, etc.) closer to a potentially affected area.
>
> Federal assistance takes many forms—including the direct provision of goods and services, financial assistance (through insurance, grants, loans, and direct payments), and technical assistance—and can come from various sources.
>
> In a major disaster or emergency as defined in the Stafford Act, the President "may direct any Federal agency, with or without reimbursement, to utilize its authorities and the resources granted to it under Federal law (including personnel, equipment, supplies, facilities, and managerial, technical and advisory services (in support of State and local assistance efforts"...

In an actual or potential Incident of National Significance that is not encompassed by the Stafford Act, the President may instruct a Federal department or agency, subject to any statutory limitations on the department or agency, to utilize the authorities and resources granted to it by Congress.

Source: National Response Plan. http://www.usda.gov/documents/NRP allpages.pdf.

It was not a normal hurricane,
and the normal disaster relief system was not equal to it

George W. Bush
SPEECH

New Orleans 17 days after Hurricane Katrina, September 15, 2005

Introduction

Hurricane Katrina, which struck the Gulf Coast in 2005, was a natural disaster equivalent to the terrorist attacks of 9/11, 2001. The hurricane overwhelmed the dikes that surrounded the city of New Orleans and left thousands of people, who had not fled the city, stranded and in peril. Levels of government faced massive coordination problems, many of which were ultimately blamed on the national government. Flood problems were compounded by the large number of toxic chemicals that were in the area. In this speech, President George W. Bush offered a message of hope, and promised to rebuild the city despite its perilous situation in a hurricane zone below sea level. Bush's speech highlights the way in which agencies designed for routine emergencies are sometimes overwhelmed by massive catastrophes and the role of multiple levels of government and private charities often play.

President Bush: Good evening. I am speaking to you from the city of New Orleans—nearly empty, still partly underwater and waiting for life and hope to return. Eastward from Lake Pontchartrain, across the Mississippi coast, to Alabama and into Florida, millions of lives were changed in a day by a cruel and wasteful storm.

In the aftermath, we have seen fellow citizens left stunned and uprooted, searching for loved ones, and grieving for the dead and looking for meaning in a tragedy that seems so blind and random.

We have also witnessed the kind of desperation no citizen of this great and generous nation should ever have to know —fellow Americans calling out for food and water, vulnerable people left at the mercy of criminals who had no mercy and the bodies of the dead lying uncovered and untended in the street.

These days of sorrow and outrage have also been marked by acts of courage and kindness that make all Americans proud. Coast Guard and other personnel rescued tens of thousands of people from flooded neighborhoods.

Religious congregations and families have welcomed strangers as brothers and sisters and neighbors. In the community of Chalmette, when two men tried to break into a home, the owner invited them to stay and took in 15 other people who had no place to go.

At Tulane Hospital for Children, doctors and nurses did not eat for days so patients could have food, and eventually carried the patients on their backs up eight flights of stairs to helicopters. Many first responders were victims themselves —wounded healers, with a sense of duty greater than their own suffering.

When I met Steve Scott of the Biloxi Fire Department, he and his colleagues were conducting a house-to-house search for survivors. Steve told me this: "I lost my house, and I lost my cars, but I still got my family, and I still got my spirit."

Across the Gulf Coast, among people who have lost much and suffered much and given to the limit of their power, we are seeing that same spirit: a core of strength that survives all hurt, a faith in God no storm can take away and a powerful American determination to clear the ruins and build better than before.

Tonight so many victims of the hurricane and the flood are far from home and friends and familiar things. You need to know that our whole nation cares about you, and in the journey ahead you are not alone. To all who carry a burden of loss, I extend the deepest sympathy of our country.

To every person who has served and sacrificed in this emergency, I offer the gratitude of our country. And tonight I also offer this pledge of the American people: Throughout the area hit by the hurricane, we will do what it takes. We will stay as long as it takes to help citizens rebuild their communities and their lives. And all who question the future of the Crescent City need to know: There is no way to imagine America without New Orleans, and this great city will rise again.

The work of rescue is largely finished. The work of recovery is moving forward. In nearly all of Mississippi, electric power has been restored. Trade is starting to return to the Port of New Orleans, and agricultural shipments are moving down the Mississippi River.

All major gasoline pipelines are now in operation, preventing the supply disruptions that many feared. The breaks in the levees have been closed, the pumps are running, and the water here in New Orleans is receding by the hour.

Environmental officials are on the ground, taking water samples, identifying and dealing with hazardous debris, and working to get drinking water and waste water treatment systems operating again.

And some very sad duties are being carried out by professionals who gather the dead, treat them with respect, and prepare them for their rest.

In the task of recovery and rebuilding, some of the hardest work is still ahead, and it will require the creative skill and generosity of a united country.

Our first commitment is to meet the immediate needs of those who had to flee their homes and leave all their possessions behind. For these Americans, every night brings uncertainty. Every day requires new courage, and the months to come will bring more than their fair share of struggles.

The Department of Homeland Security is registering evacuees who are now in shelters, churches, or private homes —whether in the Gulf region or far away. I have signed an order providing immediate assistance to people from the disaster area.

As of today, more than 500,000 evacuee families have gotten emergency help to pay for food, clothing, and other essentials. Evacuees who have not yet registered should contact FEMA or the Red Cross. We need to know who you are, because many of you will also be eligible for broader assistance in the future.

Many families were separated during the evacuation, and we are working to help you reunite. Please call this number, 1-877-568-

3317, that's 1-877-568-3317, and we will work to bring your family back together and pay for your travel to reach them.

In addition, we are taking steps to ensure that evacuees don't have to travel great distances or navigate bureaucracies to get the benefits that are there for them.

The Department of Health and Human Services has sent more than 1,500 health professionals, along with over 50 tons of medical supplies—including vaccines, antibiotics, and medicines for people with chronic conditions such as diabetes.

The Social Security Administration is delivering checks. The Department of Labor is helping displaced persons apply for temporary jobs and unemployment benefits. And the Postal Service is registering new addresses so that people can get their mail.

To carry out the first stages of the relief effort and begin the rebuilding at once, I have asked for, and the Congress has provided, more than $60 billion. This is an unprecedented response to an unprecedented crisis, which demonstrates the compassion and resolve of our nation.

Our second commitment is to help the citizens of the Gulf Coast to overcome this disaster, put their lives back together and rebuild their communities. Along this coast, for mile after mile, the wind and water swept the land clean.

In Mississippi, many thousands of houses were damaged or destroyed. In New Orleans and surrounding parishes, more than a quarter million houses are no longer safe to live in. Hundreds of thousands of people from across this region will need to find longer-term housing.

Our goal is to get people out of shelters by the middle of October. So we are providing direct assistance to evacuees that allows them to rent apartments, and many already are moving into places of their own. A number of states have taken in evacuees and shown them great compassion—admitting children to school, and providing health care. So I will work with Congress to ensure that states are reimbursed for these extra expenses.

In the disaster area and in cities that have received huge numbers of displaced people, we are beginning to bring in mobile homes and trailers for temporary use. To relieve the burden on local health care facilities in the region, we are sending extra doctors and nurses to these areas.

We are also providing money that can be used to cover overtime pay for police and fire departments while cities and towns rebuild.

Near New Orleans, Biloxi and other cities, housing is urgently needed for police and firefighters, other service providers and the many workers who are going to rebuild those cities. Right now, many are sleeping on ships we have brought to the Port of New Orleans, and more ships are on their way to the region.

And we will provide mobile homes and supply them with basic services as close to the construction areas as possible, so the rebuilding process can go forward as quickly as possible.

And the federal government will undertake a close partnership with the states of Louisiana and Mississippi, the city of New Orleans and other Gulf Coast cities, so they can rebuild in a sensible, well-planned way.

Federal funds will cover the great majority of the costs of repairing public infrastructure in the disaster zone —from roads and bridges to schools and water systems. Our goal is to get the work done quickly. And taxpayers expect this work to be done honestly and wisely, so we will have a team of inspectors general reviewing all expenditures.

In the rebuilding process, there will be many important decisions and many details to resolve, yet we are moving forward according to some clear principles. The Federal government will be fully engaged in the mission, but Gov. Barbour, Gov. Blanco, Mayor Nagin, and other state and local leaders will have the primary role in planning for their own future.

Clearly, communities will need to move decisively to change zoning laws and building codes, in order to avoid a repeat of what we have seen. And in the work of rebuilding, as many jobs as possible should go to men and women who live in Louisiana, Mississippi and Alabama.

Our third commitment is this: when communities are rebuilt, they must be even better and stronger than before the storm.

Within the Gulf region are some of the most beautiful and historic places in America.

As all of us saw on television, there is also some deep, persistent poverty in this region as well. And that poverty has roots in a history of racial discrimination, which cut off generations from the opportunity of America. We have a duty to confront this poverty with bold action.

So let us restore all that we have cherished from yesterday, and let us rise above the legacy of inequality. When the streets are rebuilt, there should be many new businesses, including minority-owned businesses, along those streets. When the houses are rebuilt, more families should own, not rent, those houses.

When the regional economy revives, local people should be prepared for the jobs being created. Americans want the Gulf Coast not just to survive, but to thrive, not just to cope, but to overcome. We want evacuees to come home, for the best of reasons —because they have a real chance at a better life in a place they love.

When one resident of this city who lost his home was asked by a reporter if he would relocate, he said, "Naw, I will rebuild, but I'll build higher." That is our vision of the future, in this city and beyond: we will not just rebuild, we will build higher and better.

To meet this goal, I will listen to good ideas from Congress, state and local officials and the private sector. I believe we should start with three initiatives that the Congress should pass.

Tonight I propose the creation of a Gulf Opportunity Zone, encompassing the region of the disaster in Louisiana and Mississippi and Alabama. Within this zone, we should provide immediate incentives for job-creating investment: tax relief for small businesses, incentives to companies that create jobs and loans and loan guarantees for small businesses, including minority-owned enterprises, to get them up and running again.

It is entrepreneurship that creates jobs and opportunity. It is entrepreneurship that helps break the cycle of poverty, and we will take the side of entrepreneurs as they lead the economic revival of the Gulf region.

I propose the creation of worker recovery accounts to help those evacuees who need extra help finding work. Under this plan, the federal government would provide accounts of up to $5,000 which these evacuees could draw upon for job training and education to help them get a good job ... and for child care expenses during their job search.

To help lower-income citizens in the hurricane region build new and better lives, I also propose that Congress pass an Urban Homesteading Act. Under this approach, we will identify property in the region owned by the federal government, and provide building sites to low-income citizens free of charge, through a lottery.

In return, they would pledge to build on the lot, with either a mortgage or help from a charitable organization like Habitat for Humanity. Home ownership is one of the great strengths of any community, and it must be a central part of our vision for the revival of this region.

In the long run, the New Orleans area has a particular challenge because much of the city lies below sea level. The people who call it home need to have reassurance that their lives will be safer in the years to come.

Protecting a city that sits lower than the water around it is not easy, but it can and has been done. City and parish officials in New Orleans, and state officials in Louisiana will have a large part in the engineering decisions to come. And the Army Corps of Engineers will work at their side to make the flood protection system stronger than it has ever been.

The work that has begun in the Gulf Coast region will be one of the largest reconstruction efforts the world has ever seen. When that job is done, all Americans will have something to be very proud of. And all Americans are needed in this common effort.

It is the armies of compassion, charities and houses of worship and idealistic men and women that give our reconstruction effort its humanity. They offer to those who hurt a friendly face, an arm around the shoulder, and the reassurance that in hard times, they can count on someone who cares.

By land, by sea and by air, good people wanting to make a difference deployed to the Gulf Coast, and they have been working around the clock ever since.

The cash needed to support the armies of compassion is great, and Americans have given generously. For example, the private fund-raising effort led by former Presidents Bush and Clinton has already received pledges of more than $100 million.

Some of that money is going to governors, to be used for immediate needs within their states. A portion will also be sent to local houses of worship to help reimburse them for the expense of helping others.

This evening the need is still urgent, and I ask the American people to continue donating to the Salvation Army, the Red Cross, other good charities and religious congregations in the region.

It is also essential for the many organizations of our country to reach out to your fellow citizens in the Gulf area. So I have asked USA Freedom Corps to create an information clearing house, available at usafreedomcorps.gov, so that families anywhere in the country can find opportunities to help families in the region or a school can support a school.

And I challenge existing organizations—churches, Scout troops, or labor union locals to get in touch with their counterparts in Mississippi, Louisiana or Alabama and learn what they can do to help. In this great national enterprise, important work can be done by everyone, and everyone should find their role and do their part.

The government of this nation will do its part as well. Our cities must have clear and up-to-date plans for responding to natural disasters, disease outbreaks or terrorist attack—for evacuating large numbers of people in an emergency and for providing the food, water and security they would need.

In a time of terror threats and weapons of mass destruction, the danger to our citizens reaches much wider than a fault line or a flood plain. I consider detailed emergency planning to be a national security priority.

Therefore, I have ordered the Department of Homeland Security to undertake an immediate review, in cooperation with local counterparts of emergency plans in every major city in America.

I also want to know all the facts about the government response to Hurricane Katrina. The storm involved a massive flood, a major supply and security operation and an evacuation order affecting more than a million people.

It was not a normal hurricane, and the normal disaster relief system was not equal to it. Many of the men and women of the Coast Guard, the Federal Emergency Management Agency, the United States military, the National Guard, Homeland Security, and state and local governments performed skillfully under the worst conditions.

Yet the system at every level of government, was not well coordinated and was overwhelmed in the first few days. It is now clear that a challenge on this scale requires greater federal authority and a broader role for the armed forces—the institution of our government most capable of massive logistical operations on a moment's notice.

Four years after the frightening experience of September 11th, Americans have every right to expect a more effective response in a time of emergency. When the federal government fails to meet such an obligation, I as president am responsible for the problem, and for the solution.

So I have ordered every Cabinet secretary to participate in a comprehensive review of the government response to the hurricane. This government will learn the lessons of Hurricane Katrina. We are going to review every action and make necessary changes so that we are better prepared for any challenge of nature, or act of evil men that could threaten our people.

The United States Congress also has an important oversight function to perform. Congress is preparing an investigation, and I will work with members of both parties to make sure this effort is thorough.

In the life of this nation, we have often been reminded that nature is an awesome force and that all life is fragile. We are the heirs of men and women who lived through those first terrible winters at Jamestown and Plymouth, who rebuilt Chicago after a great fire, and San Francisco after a great earthquake, who reclaimed the prairie from the dust bowl of the 1930s.

Every time, the people of this land have come back from fire, flood, and storm to build anew—and to build better than what we had before. Americans have never left our destiny to the whims of nature, and we will not start now.

These trials have also reminded us that we are often stronger than we know with the help of grace and one another. They remind us of a hope beyond all pain and death—a God who welcomes the lost to a house not made with hands.

And they remind us that we are tied together in this life, in this nation and that the despair of any touches us all.

I know that when you sit on the steps of a porch where a home once stood or sleep on a cot in a crowded shelter, it is hard to imagine a bright future. But that future will come.

The streets of Biloxi and Gulfport will again be filled with lovely homes and the sound of children playing. The churches of Alabama will have their broken steeples mended and their congregations whole. And here in New Orleans, the street cars will once again rumble down St. Charles, and the passionate soul of a great city will return.

In this place, there is a custom for the funerals of jazz musicians. The funeral procession parades slowly through the streets, followed by a band playing a mournful dirge as it moves to the cemetery. Once the casket has been laid in place, the band breaks into a joyful "second line"—symbolizing the triumph of the spirit over death. Tonight the Gulf Coast is still coming through the dirge, yet we will live to see the second line.

Thank you, and may God bless America.

Source: http://www.cnn.com/2005/POLITICS//09/15/bush.transcript/. Accessed 12/3/2016.

To protect the nation from all hazards, including natural disasters, acts of terrorism, and other man-made disasters

---

DEPARTMENT OF HOMELAND SECURITY APPROPRIATIONS ACT

Public Law 109-295

October 4, 2006

Introduction

Few modern catastrophes have received more media attention than Hurricane Katrina, which hit the Gulf Coast in 1985, flooding most of New Orleans and resulting in numerous deaths. The media was especially critical of the Federal Emergency Management Agency (FEMA) and its perceived failure, which many attribute to weak state and local responses, adequately to evacuate the city prior to the disaster and adequately to house them during the flood (many were housed in the Superdome). This law, adopted the following year, sought to identify the goals of FEMA and to further clarify its role in evacuations.

Much as legislation during the Cold War often focused on the threats of thermonuclear war, so laws after the 9/11 attacks of 2001 continued to emphasize the role that FEMA and other disaster agencies would play in responding to terrorist attacks.

SEC 503. FEDERAL EMERGENCY MANAGEMENT AGENCY.

(a) In general. – There is in the Department the Federal Emergency Management Agency, headed by an Administrator.

(b) MISSION—

(1) PRIMARY MISSION.—The primary mission of the Agency is to reduce the loss of life and property and protect the Nation from all hazards, including natural disasters, acts of terrorism, and other man-made disasters, by leading and supporting

the Nation in a risk-based, comprehensive emergency management system of preparedness, protection response, recovery, and mitigation.

(2) SPECIFIC ACTIVITIES.—In support of the primary mission of the Agency, the Administrator shall—

(A) Lead the Nation's efforts to prepare for, protect against, respond to, recover from, and mitigate against the risk of natural disasters, acts of terrorism and other man-made disasters including catastrophic incidents;

(B) Partner with State, local, and tribal governments and emergency response providers, with other Federal agencies, with the private sector, and with nongovernmental organizations to build a national system of emergency management that can effectively and efficiently utilize the full measure of the Nation's resources to respond to natural disasters, acts of terrorism, and other manmade disasters, including catastrophic incidents;

(C) Develop a Federal response capability that, when necessary and appropriate, can act effectively and rapidly to deliver assistance essential to saving lives or protecting or preserving property or public health and safety in a natural disaster, act of terrorism, or other man-made disaster;

(D) Integrate the Agency's emergency preparedness, protection, response, recovery, and mitigation responsibilities to confront effectively the challenges of a natural disaster, act of terrorism, or other man-made disaster;

(E) Develop and maintain robust Regional Offices that will work with State, local, and tribal governments, emergency response providers, and other appropriate entities to identify and address regional priorities;

(F) Under the leadership of the Secretary, coordinate with the Commandant of the Coast Guard, the Director of Customs and Border Protection, the Director of Immigration, and Customs Enforcement, the National Operations Center, and other agencies and offices in the Department to take full advantage of the substantial range or resources in the Department;

(G) Provide funding, training, exercises, technical assistance, planning, and other assistance to build tribal, local, State regional, and national capabilities (including communications capabilities), necessary to respond to a natural disaster, act of terrorism or other man-made disaster; and

(H) Develop and coordinate the implementation of a risk-based, all-hazrds strategy for preparedness that builds those common capabilities necessary to respond to natural disasters, acts of terrorism, and other man-made disaster while also building the unique capabilities necessary to respond to specific types of incidents that pose the greatest risk to our Nation.

SEC. 512. EVACULATION PLANS AND EXERCISES.

(a) IN GENERAL.—Notwithstanding any other provision of law, and subject to subsection (d), grants made to States or local or tribal governments by the Department through the State Homeland Security Grant Program or the Urban Area Security Initiative may be used to—

(1) Establish programs for the development and maintenance of mass evacuation plans under subsection (b) in the event of a natural disaster, act of terrorism, or other man-made disaster;

(2) Prepare for the execution of such plans, including the development of evacuation routes and the purchase and stock-piling of necessary supplies and shelters; and

(3) Conduct exercises of such plans.

(b) PLAN DEVELOPMENT.—In developing the mass evacuation plans authorized under subsection (a), each State, local, or tribal government shall, to the maximum extent practicable—

(1) establish incident command and decision making processes;

(2) ensure that State local, and tribal government plans, including evacuation routes, are coordinated and integrated;

(3) identify primary and alternative evacuation routes and methods to increase evacuation capabilities along such routes such as conversion of two-way traffic to one-way evacuation routes;

(4) identify evacuation transportation modes and capabilities, and coordinating and integrating evacuation plans for all populations including those individuals located in hospitals, nursing homes, and other institutional living facilities;

(5) develop procedures for informing the public of evacuation plans before and during an evacuation...

120 STAT. 1355.

[T]ake into account the needs of individuals with household pets and service animals

---

PETS EVACUATION AND TRANSPORTATION STANDARDS ACT OF 2006

Introduction

During natural disasters, the primary focus typically centers on the preservation of human life. During the evacuations for Hurricane Katrina, however, government officials learned that some individuals would not evacuate, even in the wake of an impending disaster, without their pets. Congress accordingly adopted amendments to the Robert T. Stafford Disaster Relief and Emergency Assistance Act to allow for government assistance to include both.

Sec. 2. Inserted the following in the earlier law.

(g) In approving standards for State and local emergency preparedness operational plans... the Director shall ensure that such plans take into account the needs of individuals with household pets and service animals prior to, during, and following a major disaster or emergency.

Sec. 3. Inserted the following:

(2) The Director may make financial contributions, on the basis of programs or projects approved by the Director, to the States and local authorities for animal emergency preparedness purposes, including the procurement, construction, leasing, or renovating of emergency shelter facilities and materials that will accommodate people with pets and service animals.

H.R. 3858 (109th): Pets Evacuation and Transportation Standards Act of 2006.

Source: https://www,.govtrack.us/congress/bills/109/hr3858/text. Accessed 11/25/2016.

[T]o improve and streamline disaster assistance for Hurricane Sandy, and for other purposes

---

DISASTER RELIEF APPROPRIATIONS ACT

2013

Introduction

In October of 2012, Hurricane Sandy made landfall in New Jersey. It resulted in 131 deaths and an estimated $30 to $50 billion in property damage in numerous northeastern states (Painter and Brown, 2013, p. 1). In early 2013, Congress funded $9.7 billion in borrowing authority for the National Flood Insurance Program (NFIP) (Id. At 2). It subsequently authorized an additional $50.7 billion in this supplemental appropriations act.

By way of background, the Budget Control Act (BCA) of 2011, which had raised the national debt ceiling while adopting measures designed to keep future budgets in check, had also provided for $11.8 billion in disaster relief for 2013. This would have left only $6.4 billion for Sandy relief (Id. At 10). This law superseded this act by evoking its emergency provision.

Although it makes sense to allocate specific amounts each year to fund disaster relief, Hurricane Sandy demonstrates that it is difficult to hold the budget line in the face of major emergencies, which may or may not remain within budget. Critics charged that the supplemental appropriation was an example of government "pork" that effectively eviscerated the Budget Control Act (Shapiro, 2014), but most politicians in affected states felt otherwise. Some critics pointed out that government paperwork delayed actual expenditures beyond the time when they might have done victims the most good.

Only a very small portion of the much larger law, which allocated monies to the Department of the Army, the Small Business Administration, the Department of Homeland Security, the Department of the Interior, the Department of Health and Human Services, the Department of Defense, the Department of Veterans Affairs, the Department of Transportation, the Department of Agriculture, and other agencies, is excerpted below.

AN ACT

Making supplemental appropriations for the fiscal year ending September 30, 2013, to improve and streamline disaster assistance for Hurricane Sandy, and for other purposes.

Sec. 1111. Recommendations for reducing costs of future disasters

(a) Report to Congress-

Not later than 1890 days after the date of enactment of this division, the Administrator of the Federal Emergency Management Agency shall submit to Congress recommendations for the development of a national strategy for reducing future costs, loss of life, and injuries associated with extreme disaster events in vulnerable areas of the United States

(b) National strategy.—

The national strategy should—

(a) respect the constitutional role and responsibilities of Federal, State, and local governments and the private sector;

(b) consider the vulnerability of the United States to damage from flooding, severe weather events, and other hazards;

(c) analyze gaps and duplication of emergency preparedness, response, recovery, and mitigation measures provided by Federal, State, and local entities; and

(d) include recommendations on how to improve the resiliency of local communities and States for the purpose of lowering future costs of disaster response and recovery.

Source: https://www.govtracki.us/congress/bills/113/hr152/text. Accessed 12/0/2016.

[T]he pessimist complains about the wind, the optimist expects it to change; the realist adjusts the sails

---

PRESIDENT BARACK OBAMA'S SPEECH TO COAST GUARD ACADEMY

May 20, 2015

Introduction

Although it is said that everyone talks about the weather but no one does anything about it, recent years have, in fact, witnessed increased attention to the phenomenon of global warming, which is being brought about by the increased release of carbon dioxide into the atmosphere by the burning of fossil fuels. Like earlier fears of the dispersion of radiation from open-air nuclear testing, this phenomenon has serious long-term impacts on the environment, including that of magnifying extremes of temperatures, droughts, flooding, and the like. In this speech, which he delivered to graduates of the U.S. Coast Guard Academy in New London, Connecticut, President Obama, whose administration had consistently stressed the need to develop renewable energy sources like wind and solar energy, tied the threats of global warming to the severity of natural disasters and stressed the need to be proactive.

... Cadets, this is a day to celebrate all that you've achieved over these past four years. You have excelled at one of the most selective and rigorous academic institutions in America. You've held yourselves to a high code of conduct, proven yourself worthy to be called commissioned officers in the United States Coast Guard.

... I want to speak to you about what comes next. Soon, you'll fan out across the Coast Guard and some of you will go to sectors and shore command. Some of you will start your duty aboard cutters. Some of you will start flight training. America needs you. And we need the Coast Guard more than ever.

We need you to safeguard our ports against all threats, including terrorism. We need you to respond in times of disaster or distress and lead your rescue teams as they jump out of perfectly good helicopters. We need you in the Caribbean and Central America, interdicting drugs before they reach our streets and damage our kids. We need you in the Middle East; in the Gulf; alongside our Navy; in places like West Africa, where you helped keep the ports open so that the world could fight a deadly disease. We need you in the Asia Pacific, to help our partners train their own coast guards to uphold maritime security and freedom of navigation in waters vital to our global economy.

These are all demanding missions. The pace of operations is intense. And these are tight fiscal times for all our services, including the Coast Guard. But we are going to keep working to give you the boats and the cutters and the aircraft that you need to complete the missions we ask of you. ...

And this brings me to the challenge I want to focus on today—one where our Coast Guardsmen are already on the front lines, and that, perhaps more than any other, will shape your entire careers—and that's the urgent need to combat and adapt to climate change.

As a nation, we face many challenges, including the grave threat of terrorism. And as Americans, we will always do everything in our power to protect our country. Yet even as we meet threats like terrorism, we cannot, and we must not, ignore a peril that can affect generations.

Now, I know there are still some folks back in Washington who refuse to admit that climate change is real. And on a day like today, it's hard to get too worried about it. There are folks who will equivocate. They'll say, "You know, I'm not a scientist." Well, I'm not either. But the best scientists in the world know that climate change is happening. Our analysts in the intelligence community know climate change is happening. Our military leaders—generals and admirals, active duty and retired—know it's happening. Our homeland security professionals know it is happening. And our Coast Guard knows it's happening.

The science is indisputable. The fossil fuels we burn release carbon dioxide, which traps heat. And the levels of carbon dioxide in the atmosphere are now higher than they have been in 800,000 years. The planet is getting warmer. Fourteen of the 15 hottest years on record have been in the past 15 years. Last year was the planet's warmest year ever recorded.

Our scientists at NASA just reported that some of the sea ice around Antarctica is breaking up even faster than expected. The world's glaciers are melting, pouring new water into the ocean. Over the past century, the world sea level rose by about eight inches. That was in the last century; by the end of this century, it's projected to rise another one to four feet.

Cadets, the threat of a changing climate cuts to the very core of your service. You've been drawn to water—like the poet who wrote, "the heart of the great ocean sends a thrilling pulse through me." You know the beauty of the sea, but you also know its unforgiving power.

Here at the Academy, climate change—understanding the science and the consequences—is part of the curriculum, and rightly so, because it will affect everything that you do in your careers. Some of you have already served in Alaska and aboard icebreakers, and you know the effects. As America's Maritime Guardian, you've pledged to remain always ready—Semper Paratus—ready for all threats. And climate change is one of those most severe threats.

And this is not just a problem for countries on the coasts, or for certain regions of the world. Climate change will impact every country on the planet. No nation is immune. So I'm here today to say that climate change constitutes a serious threat to global security, an immediate risk to our national security. And make no mistake, it will impact how our military defends our country. And so we need to act—and we need to act now.

After all, isn't that the true hallmark of leadership? When you're on deck, standing your watch, you stay vigilant. You plan for every contingency. And if you see storm clouds gathering, or dangerous shoals ahead, you don't sit back and do nothing. You take action—to protect your ship, to keep your crew safe. Anything less is negligence. It is a dereliction of duty. And so, too, with climate change. Denying it, or refusing to deal with it endangers our national security. It undermines the readiness of our forces.

It's been said of life on the sea— "the pessimist complains about the wind, the optimist expects it to change; the realist adjusts the sails." Cadets, like you, I reject pessimism. We know what we as Americans can achieve when we set ourselves to great endeavors. We are, by nature, optimists—but we're not blind optimists. We know that wishful thinking in the face of all evidence to the contrary would set us on a course for disaster. If we are to meet this threat of climate change, we must be realists. We have to readjust the sails.

That's why confronting climate change is now a key pillar of American global leadership. When I meet with leaders around the world, it's often at the top of our agenda—a core element of our diplomacy. And you are part of the first generation of officers to begin your service in a world where the effects of climate change are so clearly upon us. It will shape how every one of our services plan, operate, train, equip, and protect their infrastructure, their capabilities, today and for the long term. So let me be specific on how your generation will have to lead the way to both prepare ourselves and how to prevent the worst effects in the future.

Around the world, climate change increases the risk of instability and conflict. Rising seas are already swallowing low-lying lands, from Bangladesh to Pacific islands, forcing people from their homes. Caribbean islands and Central American coasts are vulnerable, as well. Globally, we could see a rise in climate change refugees. And I guarantee you the Coast Guard will have to respond. Elsewhere, more intense droughts will exacerbate shortages of water and food, increase competition for resources, and create the potential for mass migrations and new tensions. All of which is why the Pentagon calls climate change a "threat multiplier."

Understand, climate change did not cause the conflicts we see around the world. Yet what we also know is that severe drought helped to create the instability in Nigeria that was exploited by the terrorist group Boko Haram. It's now believed that drought and crop failures and high food prices helped fuel the early unrest in Syria, which descended into civil war in the heart of the Middle East. So, increasingly, our military and our combatant commands, our services—including the Coast Guard—will need to factor climate change into plans and operations, because you need to be ready.

Around the world, climate change will mean more extreme storms. No single weather event can be blamed solely on climate change. But Typhoon Haiyan in the Philippines gave us a possible glimpse of things to come—one of the worst cyclones ever recorded; thousands killed, many more displaced, billions of dollars in damage, and a massive international relief effort that included the United States military and its Coast Guard. So more extreme storms will mean more humanitarian missions to deliver lifesaving help. Our forces will have to be ready.

As Admiral Zukunft already mentioned, climate change means Arctic sea ice is vanishing faster than ever. By the middle of this century, Arctic summers could be essentially ice free. We're witnessing the birth of a new ocean—new sea lanes, more shipping, more exploration, more competition for the vast natural resources below.

In Alaska, we have more than 1,000 miles of Arctic coastline. The United States is an Arctic nation, and we have a great interest in making sure that the region is peaceful, that its indigenous people and environment are protected, and that its resources are managed responsibly in partnership with other nations. And that means all of you are going to have to step up—because few know the Arctic better than the U.S. Coast Guard. You've operated there across nearly 150 years. And as the Arctic opens, the role that the Coast Guard plays will only grow. I believe that our interests in the Arctic demand that we continue to invest in an enduring Coast Guard icebreaking capacity.

I was proud to nominate your last commandant, Admiral Papp, as our special representative for the Arctic. And as the U.S. chairs the Arctic Council this year, I'm committed to advancing our interests in this critical region because we have to be ready in the Arctic, as well.

Climate change, and especially rising seas, is a threat to our homeland security, our economic infrastructure, the safety and health of the American people. Already, today, in Miami and Charleston, streets now flood at high tide. Along our coasts, thousands of miles of highways and roads, railways, energy facilities are all vulnerable. It's estimated that a further increase in sea level of just one foot by the end of this century could cost our nation $200 billion.

In New York Harbor, the sea level is already a foot higher than a century ago—which was one of the reasons Superstorm Sandy put so much of lower Manhattan underwater. During Sandy, the Coast Guard mounted a heroic response, along with our National Guard and Reserve. But rising seas and stronger storms will mean more disaster response missions. And we need the Coast Guard to be ready, because you are America's maritime first responder.

Climate change poses a threat to the readiness of our forces. Many of our military installations are on the coast, including, of course, our Coast Guard stations. Around Norfolk, high tides and storms increasingly flood parts of our Navy base and an airbase. In Alaska, thawing permafrost is damaging military facilities. Out West, deeper droughts and longer wildfires could threaten training areas our troops depend on.

So politicians who say they care about military readiness ought to care about this, as well. Just as we're helping American communities prepare to deal with the impacts of climate change, we have to help our bases and ports, as well. Not just with stronger seawalls and natural barriers, but with smarter, more resilient infrastructure—because when the seas rise and storms come, we all have to be ready.

Now, everything I've discussed with you so far is about preparing for the impacts of climate change. But we need to be honest—such preparation and adaptation alone will not be enough. As men and women in uniform, you know that it can be just as important, if not more important, to prevent threats before they can cause catastrophic harm. And only way—the only way—the world is going to prevent the worst effects of climate change is to slow down the warming of the planet.

Some warming is now inevitable. But there comes a point when the worst effects will be irreversible. And time is running out. And we all know what needs to happen. It's no secret. The world has to finally start reducing its carbon emissions—now. And that's why I've committed the United States to leading the world on this challenge.

Over the past six years, we've done more than ever to reduce harmful emissions, unprecedented investments to cut energy waste in our homes and building, standards to double the fuel efficiency of our vehicles. We're using more clean energy than ever before — more solar, more wind. It's all helped us reduce our carbon emissions more than any other advanced nation. And today, we can be proud that our carbon pollution is near its lowest levels in almost two decades. But we've got to do more.

So, going forward, I've committed to doubling the pace at which we cut carbon pollution. And that means we all have to step up. And it will not be easy. It will require sacrifice, and the politics will be tough. But there is no other way. We have to make our homes and buildings more efficient. We have to invest in more energy research and renewable technologies. We have to move ahead with standards to cut the amount of carbon pollution in our power plants. And working with other nations, we have to achieve a strong global agreement this year to start reducing the total global emission—because every nation must do its part. Every nation.

So this will be tough. But as so often is the case, our men and women in uniform show us the way. They're used to sacrifice and they are used to doing hard stuff. Class of 2015, you've built new equipment that uses less energy. You've designed new vessels with fewer harmful emissions. Stephen Horvath, selected as a Fulbright Scholar, will research new technologies for renewable energies. The Coast Guard is building more fuel-efficient cutters. So you're already leading. And, Cadets, as you go forward, I challenge you to keep imagining and building the new future we need—and make your class motto your life's work: "To go where few dare." This is a place where we need you.

Across our military, our bases and ports are using more solar and wind, which helps save money that we can use to improve readiness. The Army is pursuing new, lighter, more fuel-efficient vehicles. The Air Force F-22 broke the sound barrier using biofuels. And the Navy runs an entire carrier strike group—the Green Fleet —with biofuels. Our Marines have deployed to Afghanistan with portable solar panels, lightening their load and reducing dangerous resupply missions. So fighting climate change and using energy wisely also makes our forces more nimble and more ready. And that's something that should unite us as Americans. This cannot be subject to the usual politics and the usual rhetoric. When storms gather, we get ready....

Source: Text: Obama speech on climate change at Coast Guard Academy, CT Mirror. http://ctmirror.org/23015/05/20/test-obama-speech-on-climate-change-at-coast-guard-acade... Accessed 12/3/2016.

[O]rganizations that are national in scope and purpose, voluntary, and active in disasters

---

BYLAWS

National Voluntary Organizations Active in Disasters (NVOAD)

May 22, 2016

Introduction

It is important for governmental aid to disaster areas not to depress the efforts of private charities, especially those who might be organized within or near disaster areas. The National Response Plan developed in 2004 by the Department of Homeland Security recognizes special roles for both the Red Cross and the National Voluntary Organizations in Disasters (NVOAD). The later seeks to coordinate the work of more than 40 organizations engaged in disaster relief. They include the American Red Cross, Brethren Disaster Ministries, Billy Graham Rapid Response Team, Catholic Charities USA, Feed the Children, Habitat for Humanity, the Jewish Federations, Save the Children, and Samaritan's Purse.

Article I – CORPORATION

D. PURPOSES

The purpose of the Corporation is to operate exclusively for charitable, religious, educational and scientific purposes within the meaning of Section 501(c)3 of the Internal Revenue Code...

1. To foster more effective service to people affected by disaster through communication, coordination, cooperation, collaboration, convening mechanisms and outreach;

EXHIBIT A

Membership and Partnership Categories

A. Membership Categories

1. National Membership is available to organizations that are national in scope and purpose, voluntary, and active in disasters (referred to as "National Member'). A Tier One National Member shall meet one or more of the following criteria: total organizational budget in excess of $15 million, total staff and volunteers in excess of 2,5000 or at least fifteen (15) years of prove disaster experience. A Tier Two National Member shall have either a total organizational budget in excess of $1 million or total staff and volunteers in excess of 200. Each National Members shall appoint a duly-authorized representative to represent it at meetings, serve as the primary point of contact and to be eligible for nomination to the Board of Directors. ...

2. State/Territorial Membership is available to coalitions of agencies that are statewide in scope and purpose, voluntary, and active in disasters (hereinafter referred to as "State/Territorial Member"). ...

3. Associate Membership is available to national organizations that have a vested interest throughout the disaster continuum and are able to actively contribute to the Corporation but are unable to meet the requirements of National Membership....

[T]o offer emotional and spiritual care to rescue workers and disaster survivors

---

SALVATION ARMY DISASTER RELIEF

Introduction

The Salvation Army is among the private organizations that is often lauded for its efforts during natural disasters. Its website, cited below, describes the mission of the organization as follows: "The Salvation Army, an international movement, is an evangelical part of the universal Christian Church. Its message is based on the Bible. Its ministry is motivated by the love of God. Its mission is to preach the gospel of Jesus Christ and to meet human needs in His name without discrimination."

Originally established in England by William Booth (1829-1912), a Methodist minister, the Salvation Army was known for reaching to the down-and-out, especially in large cities. It continues to devote a significant part of its resources for caring for the homeless. Although it has a distinct evangelical emphasis, it has a nondiscrimination policy with regard to its services. Unlike secular governments, the Salvation Army can minister to spiritual as well as physical needs.

Disaster Relief Services

While each disaster creates its own unique circumstances and special needs, Salvation Army disaster relief efforts focus on seven core services. These services may be modified based on the magnitude of the disaster and adapted to meet the specific needs of individual survivors.

Training

The first step in being ready to respond to an emergency is training. In partnership with other agencies, The Salvation Army's disaster training program offers a variety of courses designed to help individuals and communities prepare for emergency events and become trained disaster volunteers.

Food Service

When disaster strikes, one of the first signs that help is on the way is often the arrival of a Salvation Army mobile feeding unit, offering meals, snacks and drinks to rescue workers and survivors.

Emotional and Spiritual Care

Motivated by Christian faith, The Salvation Army deploys specially trained individuals to offer emotional and spiritual care to rescue workers and disaster survivors.

Emergency Communications

The Salvation Army Team Emergency Radio Network (SATERN) includes a worldwide network of volunteer amateur radio operators and other communications specialists, who may be mobilized to transmit emergency information during a disaster event.

Disaster Social Services

After a family has lost everything in a disaster, The Salvation Army is there to provide emergency assistance to help meet survivors' most urgent needs for food, clothing, shelter and medical services.

Donations Management

The Salvation Army is one of the nation's leaders in responsibly collecting, sorting and distributing donated goods. The Salvation Army encourages cash donations as the best and most flexible way to help and solicits only those in-kind donations which can be effectively received and efficiently distributed.

Recovery

The Salvation Army supports long-term disaster recovery operations with flexible programming that is adaptable to the unique needs of individual communities.

Source: Taken from Salvation Army home page at https://disaster.salvationarmyusa.org/aboutus/?ourservices Accessed 11/15/2016.

Significantly improved the detection and observation of environmental phenomena

---

STATEMENT ON LAUNCH OF GOES-R SATELLITE

November 19, 2016

Introduction

Probably few governmental functions are more taken for granted than that of weather forecasting. Such forecasts allow cities and localities to get prior warnings of impending tornadoes, thunder-storms, hurricanes, and tsunamis so they can escape or seek shelter. On November 19, 2016, the U.S. launched a sophisticated new satellite known as the GOES-R Satellite. The information from this and other satellites coordinated by the National Oceanic and Atmospheric Administration (which is part of the U.S. Department of Commerce), the National Aeronautics and Space Administration (NASA), and the Department of Defense, is shared with approximately 200 other governments through the World Meteorological Organization.

(Washington, DC) – Today, the Geostationary Operational Environmental Satellite - R Series (GOES-R) satellite launched successfully from Cape Canaveral Air Force Station. Once in geostationary orbit, GOES-R will be known as GOES-16. The GOES-R program represents the nation's next generation of geostationary weather satellites. Its development has been managed by NOAA's National Environmental Satellite, Data, and Information Services (NESDIS) in partnership with the National Aeronautics and Space Administration (NASA). The data collected by this satellite will provide critical inputs to NOAA's weather prediction capabilities and support weather forecasting activities around the globe.

Ranking Member Eddie Bernice Johnson (D-TX) made the following statement.

> "I want to congratulate everyone involved in making this a successful launch: NOAA, NASA, and the entire contractor team. GOES-R will be an essential element of our weather forecasting fleet, and, as NOAA has stated, will 'significantly improve the detection and observation of environmental phenomena that directly affect public safety, protection of property and our nation's economic health and prosperity.'"

Source: http://democrats.science.house.gov/press-release/ranking-member-johnson-statement-suc... 11/20/2016.

See Angela Fritz, "U.S. launches next-generation weather satellite that will revolutionize forecasting," https://www.washingtonpost.com/news/capital-weather-gang/wp/2016/11/19/u-s-launche... Accessed 11/20/2016.

Americans have always come to to aid of their fellow countrymen

---

PRESIDENT DONALD J. TRUMP PROCLAIMS SEPTEMBER 3, 2017, AS A NATIONAL DAY OF PRAYER FOR THE VICTIMS OF HURRICANE HARVEY AND FOR OUR NATIONAL RESPONSE AND RECOVERY EFFORTS

Introduction

Hurricanes remain a recurring national threat, especially to the Gulf Coast region. In August 2017, Texas was hit by a massive storm named Harvey that set a record for the most rainfall recorded in a single such event. Although such rains would likely cause damage anywhere, their effects were arguably exacerbated by the rapid development of the city of Houston, the nation's fourth largest. Although the death toll was kept to just over 50, first responders and volunteers (including the so-called Cajun Navy from Louisiana) made thousands of rescues many from homes in high-end neighborhoods that had never flooded before. Going back to a time-honored tradition, President Trump issued the following proclamation of a National Day of Prayer. Trump made two visits to the area immediately during and after the storm. Congress subsequently adopted a $15.3 billion relief measure, which was tied to raising the national debt ceiling, just as Hurricane Irma, a massive Category 5 storm, was, after ravaging a number of Caribbean islands, was taking aim for Florida.

A Proclamation

Hurricane Harvey first made landfall as a Category 4 storm near Rockport, Texas, on the evening of August 25, 2017. The storm has since devastated communities in both Texas and Louisiana, claiming many lives, inflicting countless injuries, destroying or damaging tens of thousands of homes, and causing billions of dollars in damage. The entire Nation grieves with Texas and Louisiana. We are deeply grateful for those performing acts of service, and we pray for healing and comfort for those in need.

Americans have always come to the aid of their fellow countrymen—friend helping friend, neighbor helping neighbor, and stranger helping stranger—and we vow to do so in response to Hurricane Harvey. From the beginning of our Nation, Americans have joined together in prayer during times of great need, to ask for God's blessings and guidance. This tradition dates to June 12, 1775, when the Continental Congress proclaimed a day of prayer following the Battles of Lexington and Concord, and April 30, 1789, when President George Washington, during the Nation's first Presidential inauguration, asked Americans to pray for God's protection and favor.

When we look across Texas and Louisiana, we see the American spirit of service embodied by countless men and women. Brave first responders have rescued those stranded in drowning cars and rising water. Families have given food and shelter to those in need. Houses of worship have organized efforts to clean up communities and repair damaged homes. Individuals of every background are striving for the same goal —to aid and comfort people facing devastating losses. As Americans, we know that no challenge is too great for us to overcome.

As response and recovery efforts continue, and as Americans provide much needed relief to the people of Texas and Louisiana, we are reminded of Scripture's promise that "God is our refuge and strength, a very present help in trouble." Melania and I are grateful to everyone devoting time, effort, and resources to the ongoing response, recovery, and rebuilding efforts. We invite all Americans to join us as we continue to pray for those who have lost family members or friends, and for those who are suffering in this time of crisis.

NOW, THEREFORE, I, DONALD J. TRUMP, President of the United States of America, do hereby proclaim September 3, 2017, as a National Day of Prayer for the Victims of Hurricane

Harvey and for our National Response and Recovery Efforts. We give thanks for the generosity and goodness of all those who have responded to the needs of their fellow Americans. I urge Americans of all faiths and religious traditions and backgrounds to offer prayers today for all those harmed by Hurricane Harvey, including people who have lost family members or been injured, those who have lost homes or other property, and our first responders, law enforcement officers, military personnel, and medical professionals leading the response and recovery efforts. Each of us, in our own way, may call upon our God for strength and comfort during this difficult time. I call on all Americans and houses of worship throughout the Nation to join in one voice of prayer, as we seek to uplift one another and assist those suffering from the consequences of this terrible storm.

IN WITNESS WHEREOF, I have hereunto set my hand this first day of September, in the year of our Lord two thousand seventeen, and of the Independence of the United States of America the two hundred and forty-second.

DONALD J. TRUMP

Source: https://www.whitehouse.gov/the-press-office/2017/09/01/president-donald-j-trump-proclaims-september-3-2017-national-day-prayer. Accessed 9/5/1017.

@POTUS has authorized the Jones Act be waived

TWEET WAIVING JONES ACT

2017

Introduction

The 2017 hurricane season proved to be one of the most devastating in U.S. history. Hurricane Harvey attained category intensity before hitting Texas and Louisiana starting on August 23. It deposited more than 50 inches of rain in parts of Texas and resulted in substantial flooding in the city of Houston. Hurricane Irma, one of the strongest every recorded, subsequently hit Florida on September 10 and wrecked devastation throughout much of the state. Hurricane Maria subsequently hit Puerto Rica, which had already been damaged by the prior hurricanes, with category 4 intensity on September 20, 2017.

The latter hurricane took out all public power on the island and caused massive devastation to the islands infrastructure with about three and a half million inhabitants, most of whom are U.S. citizens. The island was already on the verge of bankruptcy, and federal efforts, perhaps because they were already stretched thin by the two previous storms, were strongly criticized.

The Merchant Marine Act of 1920, which is more generally called the Jones Act, is a protectionist measure designed to protect the U.S. Merchant Marine by prohibiting transport between U.S. ports by ships not registered and constructed in the U.S. and owned by U.S. citizens. Foreign ships seeking to transport goods to Alaska, Hawaii, Puerto Rico, or other U.S. islands must unload their goods on the mainland for shipment on U.S. ships. This increases the costs involved as U.S. shipping are traditionally more expensive.

After noting that the President had issued an executive order to waive this requirement for states affected by Hurricanes Harvey and Irma, Sarah Huckabee Sanders, President Donald Trump's Press Secretary, issued the following tweet on behalf of the President of the United States (POTUS), to which Puerto Rican Governor Ricardo Rosello responded.

Sarah Sanders
@PrsssSec.
At @ricdorossello request, @POTUS has
Authorized the Jones Act be waived for Puerto Rico.
It will go into effect immediately.
7:02 AM – Sep 28, 2017.

Richardo Rossello
@ricardorossello
Thanks you @POTUS twitter.com/presssec/statu…
7:10 AM – Sep 28, 2017

Source: Niraj Chokshi, September 28, 2017. *The New York Times*. "Trump Waives Jones Act for Puerto Rico, Easing Hurricane Aid Shipments, https://www.nytimes.com/2017/09/28/us/jones-act-waived.html?mcubz=1&_r=0

I know it is really a good news story

VERBAL EXCHANGE BETWEEN HOMELAND SECURITY SECRETARY ELAINE DUKE AND SAN JUAN MAYOR CARMEN YULIN CRUZ

September 28-29, 2017

Introduction

It is often possible to view catastrophes from different angles. After Puerto Rico, an American territory, was almost totally devastated by Hurricane Maria in September 2017, American officials could point to the relatively few people who lost their lives, but those on the ground on that island faced power outages, disruption of telephone services, gas shortages, and obstructed roads.

As the following excerpts show, on September 28, Elaine Duke, the acting Secretary of Homeland Security tried to put a positive spin on the situation. The next day, San Juan Mayor Carme Yulin Cruz responded with a sense of exasperation. Much of the island remained out of power well into the new year. A study published in *The New England Journal of Medicine* entitled "Mortality in Puerto Rico after Hurricane Maria" in May 2018 by a group of researchers from Harvard University indicated that the actual death count from the aftermath of Hurricane Maria in Puerto Rico is likely "more than 70 times the official estimate" of 64.

Duke's comments:

"I know it is really a good news story in terms of our ability to reach people and the limited number of deaths that have taken place in such a devastating hurricane."

Mayor Carmen's response to hearing this clip in an interview with CNN's Alisyn Camerota:

"Well maybe from where she's standing it's a good news story. When you're drinking from a creek, it's not a good news story. When you don't have food for a baby, it's not a good news story. When you have to pull people down from buildings—I'm sorry, that really upsets me and frustrates me. I would ask you [Duke] to come down here and visit the towns and then make a statement like that, because frankly, it is an irresponsible statement in contrast with the statements of support that I have been getting yesterday when I got that call from the White House. Dammit, this is not a good news story. This is a 'people are dying' story. It's a life-or-death story."

Source: Taken from Daniella Diaz, "San Juan Major: 'Dammit, this is not a good news story." CNN. http://www.cnn/com/2017/09/29/puerto-rico-hurricane-maria-san-juan-mayor-trump-response/index.html.

Not to exclude houses of worship from eligibility

REVISED FEMA REGULATIONS

(January 2018)

Introduction

Observers have long noted that lightning does not exempt houses of worship from its destruction. Neither do other natural disasters. Because the First Amendment to the U.S. Constitution prohibits an "establishment" of religion, the Federal Emergency Management Agency (FEMA) has exempted houses of worship from most disaster relief.

A number of churches challenged this policy in the wake of hurricanes that struck Texas, Florida, and Puerto Rico in 2017, arguing that the policy appeared to infringe on the "free exercise of religion (also guaranteed by the First Amendment) by discriminating against religious institutions. In June 2017, the U.S. Supreme Court had issued a decision in which it had ruled that a state could not arbitrarily deny aid to a playground simply because it was operated by a church. Taking a cue from this decision, FEMA reviewed its policies in January of 2018 to permit such aid. Churches had already filed suits against prior policies, which (if not now considered moot), might still be ultimately be decided by the courts.

Foreword to Public Assistance Program and Policy Guide
January 2018

On June 26, 2017, the United States Supreme Court held in *Trinity Lutheran Church of Columbia v. Comer,* that the Missouri Department of Natural Resources violated the Free Exercise Caluse of the First Amendment of the United States Constitution by denying Trinity Lutheran Church an otherwise available public benefit solely because it was a church. In light of the *Trinity Lutheran* decision, FEMA has considered its guidance of private nonprofit facilitiy eligibility and determined that it will revise its interpretation of the aforementioned statutory and regulatory authorities so as not to exclude houses of worship from eligibility for REMA aid on the basis of the religious character or primarily religious use of the facility."

Source: See *Public Assistance Program and Policy Guide* – EF-104-009-2/ January 2018. Washington, D.C.: FEMA, p. vii. Available on the Internet.

A Historic Year of Weather and Climate Disasters

NATIONAL CENTERS FOR ENVIRONMENTAL INFORMATION (NCEI) REPORT ON 2017 DISASTERS

(January 2018)

Introduction

Although every year brings its own natural catastrophes, some years are worse than others. Climate scientists, especially those concerned about global warming, are typically reluctant to trace any specific event to this general trend. Nonetheless, they expect that this phenomenon will bring about greater extremes in weather events. Recent data compiled by the NOAA National Centers for Environmental Information (NCEI) seem to confirm that both the number and severity of events have been increasing over the past 5 years.

2017 in Context

In 2017, there were 16 weather and climate disaster events with losses exceeding $1 billion each across the United States. These events included 1 drought event, 2 flooding events, 1 freeze event, 8 severe storm events, 3 tropical cyclone events, and 1 wildfire event. Overall, these events resulted in the deaths of 362 people and had significant economic effects on the areas impacted. The 1980-2017 annual average is 5.8 events (CPI-adjusted); the annual average for the most recent 5 years (2013-2017) is 11.6 events (CPI-adjusted).

During 2017, the U.S. experienced a historic year of weather and climate disasters. In total, the U.S. was impacted by 16 separate billion-dollar disaster events typing 2011 for the record number of billion-dollar disasters for an entire calendar year. In fact, 2017 arguably has more events than 2011 given that our analysis traditionally counts all U.S. billion-dollar wildfires, as regional-scale, seasonal events, not as multiple isolated events.

More notable than the high frequency of these events is the cumulative cost, which exceeds $300 billion in 2017 – a new U.S. annual record. The cumulative damage of these 16 U.S. events during 2017 is $306.2 billion, which shatters the previous U.S. annual record cost of $214.8 billion (CPI-adjusted), established in 2005 due to the impacts of Hurricanes Dennis, Katrina, Rita and Wilma.

Source: NOAA National Centers for Environmental Information (NCEI) U.S. Billion-Dollar Weather and Climate Disasters (2018). https://www.ncdc.noaa.gov/billions/

Glossary

Acts of God A term, common in the insurance industry, to describe events that are beyond human control. In reality, human efforts at prevention – for example, installation of lightning rods—can sometimes prevent or mitigate such disasters.

Catastrophic incident Used to designate incidents of national significance, which calls for highest levels of aid and intergovernmental cooperation.

Civil Defense Programs and efforts designed to protect citizens from military or natural disasters.

Department of Homeland Security The governmental agency, created after the 9/11 terrorist attacks, tasked both with protecting the nation and with responding to natural disasters. It contains the Federal Emergency Management Agency (FEMA).

Disaster Calamitous event causing great destruction and loss. Such events may be either natural or man-made, with human actions often exacerbating natural phenomena.

DRO Acronym for a disaster relief organization.

Emergency Declarations Presidential declarations designed to provide aid to state and local communities, sometimes for impending disasters.

Eminent Domain The power of governments to take property for public purposes. The Fifth Amendment of the Constitution requires the government to provide "just compensation" in such circumstances.

Enumerated Powers Powers that are specifically listed within the Constitution. These may be supplemented with implied powers.

Euergetism The practice of senators, emperors and other leading political figures in Greek and Roman history to provide public entertainment and to construct magnificent building for their constituents, especially in the aftermath of natural disasters.

Faith-based Organizations Those created and administered by individuals who are motivated by religious sentiments and obligations. They have traditionally played an important role in humanitarian disaster relief.

Federalism A system, like that in the United States, that divides governmental responsibilities between the nation and the states and their localities. Most national laws related to natural disasters assume that state and local officials will serve as first responders.

FEMA Federal Emergency Management Agency was created by an executive order in 1978 and now operates under the U.S. Department of Homeland Security.

Fire Management Assistance Grant Declarations May be issued either by the President or a FEMA Regional Director and provides equipment, personnel and other aid to prevent a fire or series of fires from becoming a larger disaster. Such declarations do not provide direct aid to individuals or households.

First Responders Those, typically state and local officials, who are first to arrive on the scene of disasters.

Free Rider Problem Occurs when an individual or government reduce their own efforts in mitigating potential disasters in anticipation that another (typically, the national government) will do so.

Laissez-faire The idea that private markets and individual efforts will address major problems of economic distribution and that governmental intervention, other than to assure competition, often results in the misallocation of resources.

Local responders Local leaders who often take the lead in responding to emergencies.

Mitigation Efforts aimed at reducing the impact of disaster, often by taking preventive measures in advance.

Moral Hazard The way in which promises of governmental disaster relief might undercut the caution that individuals might otherwise exercise in preparing for future disasters.

National Guard Military units (successors to militia) under the control of the states but which the president can sometimes federalize in times of emergencies.

New Deal Name for programs proposed by President Franklin D. Roosevelt to combat the Great Depression. They called for a greatly expanded role for the national government.

NGOs Nongovernmental organizations, who are often key participants in disaster relief.

Paternalism The idea that government, rather than private individuals, should assume primary responsibility for reacting to disasters.

Police Powers Authority over issues like land use, policing, and education that are primarily reserved to the states through the Tenth Amendment to the U.S. Constitution.

Posse Comitatus Group of private citizens enlisted for law enforcement. The Posse Comitatus Act of 1878 limits the use of the U.S. military in civilian operations.

Preliminary Damage Assessments (PDAs) Compiled by Federal Emergency Management Agency (FEMA) as a means of providing information as to whether governors should request federal disaster declarations.

Preparedness The degree to which individuals, organizations, and governments are ready for disasters.

Prevention Efforts to prepare in advance to mitigate the effects of natural disasters.

Presidential Declaration of a Federal Major Disaster or Emergency. Opens the doors for increased federal aid in the aftermath of such events.

Recovery Used both to describe efforts to obtain the bodies of victims who have died and to build infrastructures that were damaged by disasters.

Relief Aid that is extended in the immediate aftermath of natural disasters.

Rescue Efforts to save individuals who are caught in harm's way, as in flood waters, fires, or landslides.

Safety and security The former term is generally used to describe protection from all dangers, including natural disasters, whereas the latter term generally describes attempts to protect against dangers from malevolent individuals or enemy nations.

State of Nature Hypothetical state envisioned by social contract philosophers to explain what life would be like if there were no governmental institutions. Extreme disaster situations can sometimes return individuals to a similar state.

Taxing and Spending Clause A provision in Article I, Section 8 of the Constitution granting congress to tax and spend on behalf of the general welfare and often used as the constitutional justification for governmental disaster relief.

Terrorism The use of violence against civilians often by non-state actors for religious, political, or ideological goals.

Turndowns Presidential denials of gubernatorial requests for disaster declarations.

Zoning Designating certain areas for certain kinds of building and residences. Zoning can be used to mitigate possible future disasters.

Bibliography

Ackerman, Bruce. 2004. "The Emergency Constitution." 113 *Yale Law Journal* (March): 1029-1091.

Alexander, David. 2001. *Natural Disasters.* New York: Routledge.

American Red Cross. "Our Federal Charter." http://www.redcross.org/about-us-history/federal-charter. Accessed 10/1/2016.

Baca, Anna Marie. September 2008. "History of Disaster Legislation." *On Call* (vol. 1, Issue 3). FEMA publication, pp. 1-3.

Barry, John M. 1998. *Rising Tide: The Great Mississippi Flood of 1927 AND How it Changed America.* New York: Simon & Schuster.

Ben-Shahar, Omri and Kyle D. Logue. 2016. "The Perverse Effects of Subsidized Weather Insurance." *Stanford Law Review* 68 (March): 571-626.

Birkland, Thomas A. 2006. *Lessons of Disaster.* Washington, D.C.: Georgetown University Press.

Birkland, Thomas and Sarah Waterman. 2008. "Is Federalism the Reason for Policy Failure in Hurricane Katrina?" *Publius* 38 (Fall): 692-714.

Brinkley, Douglas. 2006. *The Great Deluge: Hurricane Katrina, New Orleans, and the Mississippi Gulf Coast.* New York: Harper Collins.

Burton, Ian, Robert W. Kates, and Gilbert F. White. 1993. *The Environment as Hazard,* 2nd ed. New York: Guilford Press.

Caldwell, Shirley W. 2003. "'God Help Them All and so Must We': Clara Barton, Reverend John Brown, and Drought Relief Efforts, 1886-1887." *The Southwestern Historical Quarterly* 106 (April): 507-530.

Carafano, James Jay and Richard Weitz. March 21, 2006. "Learning from Disaster: The Role of Federalism and the Importance of Grassroots Response." http;//www.heritage.org/research/reports/2006/03/learning-from-disaster-the-role-of-feder... 11/10/2016.

Cords, Danshera. 2015. "Charity Begins at Home? An Exploration of the Systemic Distortions Resulting from Post-Disaster Giving Incentives," *Rutgers Law Journal* 44 (Winter): 213-270.

Dallek, Matthew. 2016. *Defenseless Under the Night: The Roosevelt Years and the Origins of Homeland Security*. New York: Oxford University Press.

Dauber, Michele Landis. 2013. *The Sympathetic State: Disaster Relief and the Origins of the American Welfare State*. Chicago: University of Chicago Press.

Dickson, Charles Ellis. 1987. "Jeremiads in the New American Republic: The Case of National Fasts in the John Adams Administration." *The New England Quarterly* 60 (June): 187-207.

Elsea, Jennifer K. September 16, 2005. *The Use of Federal Troops for Disaster Assistance: Legal Issues*, CRS Report for Congress, Received through the CRS Web.

Ferris, Elizabeth G. 2011. *The Politics of Protection: The Limits of Humanitarian Action*. Washington, D.C.: Brookings Institution.

Fountain, Philip M., Sara L. Kindon, and Warwick E. Murray. 2004. "Christianity, Calamity, and Culture: The Involvement of Christian Churches in in the 1998 Aitape Tsunami Disaster Relief." *The Contemporary Pacific* 17 (Fall): 321-355.

Fritz, Angela. November 19, 2016. "U.S. launches next-generation weather satellite that will revolutionize forecasting." https://www.washingtonpost.com/news/capital-weather-gang/wp/2016/11.19.u-s-launche... 11/29/2016.

Grada, Cormac O. and Kevin H. O'Rourke. 1997. "Migration as disaster relief: Lessons from the Great Irish Famine." *European Review of Economic History* 2 (April) 3-25.

Gunn, Angus M. 2008. *Encyclopedia of Disasters: Environmental Catastrophes and Human Tragedies*, 2 vols. Westport, CT: Greenwood Press.

Guyton, Gregory P. 1999. "A Brief History of Workers' Compensation." *Iowa Orthopedia Journal* 19:106-110. http;//europepmc.org/articles/PMC1888620. Accessed 12/13/2016.

Haraldsson, Hrafnkell, "Phil Robertson Dishonestly Claims Founding Fathers Opposed Federal Disaster Relief," September 9,

2016. Found at http://www.politicususa.com/2016/09/09/phil-robertson-dishonestly-claims-founding-father. Accessed 12/5/2016.

Heath, J. Benton. 2011. "Disasters, Relief, and Neglect: The Duty to Accept Humanitarian Assistance and the Work of the International Law Commission." *International Law and Politics* 43: 419-477.

Higgins, Christopher M. *Popular and Imperial Response to Earthquakes in the Roman Empire.* M.A. thesis at Ohio University, June, 2009.

Hochschild, Arlie Russell. 2016. *Strangers in Their Own Land: Anger and Mourning on the American Right.* New York: The New Press.

Hokanson, Petra. 6/1/2016. "American Destruction: How the Swamp Land Act of 1850 Wreaked Havoc on Florida's Everglades." http://culturalhistoryus.weebly.com/american-cultural-history/-american-destruction-how-the-swamp-land-act-of-1850-wreaked-havoc-on-floridas-everglades-by-petra-hokanson.

Kent, Elizabeth F. 2006. "'Where's the Calvary?' Federal Response to 21st Century Disasters 40 *Suffolk University Law Review* 181-213.

Kosar, Kevin R. March 15, 2006. *The Congressional Charter of the American Red Cross: Overview, History, and Analysis.* Congressional Research Service: The Library of Congress.

Landis, Michele L. 1999. "Fate, Responsibility, and 'Natural' Disaster Relief: Narrating the American Welfare State," 33 *Law & Society Review* 257-313.

Lindsay, Bruce R. and Francis X. McCarthy. June 14, 2015. *Stafford Act Declarations 1953-2015: Trends, Analyses, and Implications for Congress.* Congressional Research Service.

McCullough, David. 1968. *The Johnstown Flood: The Incredible Story Behind One of the Most Devastating Disasters America Has Ever Known.* New York: Simon & Schuster.

McMillan, Christine M. 2007. "Federal Flood Insurance Policy: Making Matters Worse." *Houston Law Review* 44 (Summer): 471-504.

Mason, Alpheus T. and Gordon E. Baker. 1985. *Free Government in the Making: Readings in American Political Thought*, 4th ed. New York: Oxford University Press.

Mener, Andrew S. "Disaster Response in the United States of America: An Analysis of the Bureaucratic and Political History of a Failing System." 10 May 2007. CUREJ: College Undergraduate Research Electronic Journal, University of Pennsylvania, http://repository.upenn.edu/curej/63.

Miskel, James F. 2008. *Disaster Response and Homeland Security: What Works, What Doesn't.* Stanford, CA: Stanford Security Studies.

Mulcahy, Matthew. 2006. *Hurricanes and Society in the British Greater Caribbean, 1624-1783.* Baltimore: The John Hopkins University Press.

Mulcahy, Matthew. 1998. "The 'Great Fire' of 1740 and the Politics of Disaster Relief in Colonial Charleston. *The South Carolina Historical Magazine* 99 (April): 135-157.

National Response Framework. May 2013. 2nd ed. Department of Homeland Security.

Nuttli, Otto W, G.A. Bollinger, and Robert B. Herrmann. 1986. *The 1886 Charleston, South Carolina, Earthquake – A 1986 Perspective.* U.S. Geological Survey Circular 985. U.S. Government Printing Office.

Olasky, Marvin. 2006. *The Politics of Disaster: Katrina, Big Government, and a New Strategy for future Crises.* Nashville, TN: W Publishing Group.

Painter, William L., and Jared T. Brown. February 19, 2013. *FY2013 Supplemental Funding for Disaster Relief.* Congressional Research Services.

Percy, William Alexander. 1966. *Lanterns on the Levee: Recollections of a Planter's Son.* New York: Alfred A. Knopf.

Platt, Rutherford H. 1999. *Disasters and Democracy: The Politics of Extreme Natural Events.* Washington, D.C.: Island Press.

Reale, Andrena, September 2010 "Acts of God(s); the role of religion in Disaster risk Reduction," http://odihpn.org/magazine/a

cts-of-gods-the-role-of-religion-in-disasterrisk-reduction. Accessed 9/28/2016.

Roberts, Patrick S. 2013. *Disasters and the American State: How Politicians, Bureaucrats, and the Public Prepare for the Unexpected.* 2nd ed. New York: Cambridge University Press.

Sarat, Austin and Javier Lezaun. 2009. *Catastrophe: Law, Politics, and the Humanitarian Impulse.* Amherst: University of Massachusetts Press.

Schneider, Saundra. 2011. *Dealing with Disaster: Public Management in Crisis Situations.* 2nd ed. Armonk, New York: M.E. Sharpe.

Shapiro, Philip O. 2014. "A Sustainable Budget Should Endure Any Storm," *New York University Journal of Legislation and Public Policy* 17: 595-635.

Shugerman, Jed Handelsman. 2000. "The Floodgates of Strict Liability: Bursting Reservoirs and the Adoption of *Fletcher v. Rylands* in the Gilded Age," *Yale Law Journal* 110 (January): 333-377.

Shughart, William F., II. 2011. "Disaster Relief as Bad Public Policy." *The Independent Review* 15 (Spring): 519-539.

Special Regulations No. 67, 1917. *Regulations Governing Flood Relief Work of the War Department.* Washington: Government Printing Office.

Steinberg Ted. 2000. *Acts of God: The Unnatural History of Natural Disaster in America.* New York: Oxford University Press.

Sterett, Susan. 1997. "Serving the State: Constitutionalism and Social Spending, 1860s-1920s," *Law & Social Inquiry* 22 (Spring): 311-356.

Sun, Lisa Grow. "Disaster Mythology and the Law." *Cornell Law Review* 95 (July): 1131-1207.

Wright, James M. April 2000. *The Nation's Responses to Flood Disasters: A Historical Account.* A Report by the Association of State Floodplain Managers, ed. By Wender L. Hessler.

Thorndike, Joseph J. December 13, 2012. "How the Charity Deduction Made the World Safe for Philanthropy."

http://www.taxhistory.org/thp/readings.nsf/ArtWeb/972168BEA0B68D8585257B160048... 12/12/2016.

Wright, Jonathan B. and Douglas A. Hicks, "Disaster relief: What would Adam Smith do?" *CSMonitor*. http://www.csmonitor.com/2005/0118/p09s02-coop.html. Accessed 12/6/2016.

Veyne, Paul. 1990. *Bread and Circuses: Historical Sociology and Political Pluralism*. Abridged by Oswyn Murray, trans. By Brian Pearce. New York: Penguin Books.

Vile, John R. 2015. *A Companion to the United States Constitution and Its Amendments*, 6th ed. Santa Barbara: Praeger.

______. 2016. *The Constitutional Convention of 1787: A Comprehensive Encyclopedia of America's Founding*, Rev. 2nd ed. 2 vols. Clark, NJ: Talbot.

_____. 2016. *American Immigration and Citizenship: A Documentary History*. Lanham, MD: Rowman and Littlefield.

Warren, John K. 2007. "Restoring Responsibility and Accountability in Disaster Relief," 31 *William and Mary Environmental Law and Policy Review* (Spring): 893-925.

Wilkavsky, Aaron. 1984. *The Nursing Father: Moses as a Political Leader*. University of Alabama Press.

Wilkins, Dexter. "A Bigger Problem Than ISIS?" January 2, 2017, *The New Yorker*. http://www.newyorker.com/;magazine/2017/01/02/a-bigger-problem-than-isis.

Wright, J.O. 1907. *Swamp and Overflowed Lands in the United States. Ownership and Reclamation*. Washington: Government Printing Office for U.S. Department of Agriculture.

Zack, Naomi. 2009. *Ethics for Disaster*. Lanham, MD: Rowman & Littlefield.

Zack, Naomi, "Philosophy and Disaster," *Homeland Security Affairs* vol. 2006, No. I, Article 5. Accessible at http://www.hsaj.org/hsa/volII/iss1/art5.

Index

About the Author

John R. Vile (Ph.D., UVA) is Dean of the Honors College at Middle Tennessee State University. His recent books include *The Constitutional Convention of 1787: A Comprehensive Encyclopedia of America's Founding*, rev. 2nd ed. (2016), *A Companion to the United States Constitution* (6th ed., 2015); *Encyclopedia of Constitutional Amendments, Proposed Amendments, and Amending Issues* (4th ed., 2015); *American Immigration and Citizenship: A Documentary History* (2016); *Essential Supreme Court Decisions: Summaries of Leading Cases in U.S. Constitutional Law* (17th ed., 2018); and a series of four volumes (2015-2018) decoding documents from the American Founding Period through the Civil War and Reconstruction periods.